DISRUPTIVE TECHNOLOGIES, CLIMATE CHANGE AND SHIPPING

MARITIME AND TRANSPORT LAW LIBRARY

Ship Building, Sale and Finance
Edited by Barış Soyer and Andrew Tettenborn

The Modern Law of Marine Insurance
Volume 4
Edited by D. Rhidian Thomas

Air Cargo Insurance
Malcom A. Clarke and George Leloudas

Offshore Oil and Gas Installations Security
An International Perspective
Mikhail Kashubsky

International Trade and Carriage of Goods
Edited by Barış Soyer and Andrew Tettenborn

Maritime Law and Practice in China
Liang Zhao and Lianjun Li

Maritime Cross-Border Insolvency
Lia Athanassiou

The Law of Yachts and Yachting
Second Edition
Edited by Filippo Lorenzon and Richard Coles

Maritime Liabilities in a Global and Regional Context
Edited by Barış Soyer and Andrew Tettenborn

New Technologies, Artificial Intelligence and Shipping Law in the 21st Century
Edited by Barış Soyer and Andrew Tettenborn

The Law of Wreck
Nicholas Gaskell and Craig Forrest

Codification of Maritime Law
Challenges, Possibilities and Experience
Edited by Zuzanna Pepłowska-Dąbrowska and Justyna Nawrot

Transport Documents in Carriage of Goods by Sea
International Law and Practice
By Caslav Pejovic

Maritime Law
Fifth Edition
Edited by Yvonne Baatz

Ship Operations: New Risks, Liabilities and Technologies in the Maritime Sector
Edited by Barış Soyer and Andrew Tettenborn

Maritime Safety in Europe
Edited by Justyna Nawrot and Zuzanna Pepłowska-Dąbrowska

Fresh Produce Shipping
Damages and Compensation
Rex C. Tester

Freeports and Free Zones
Operations and Regulation in the Global Economy
Mark Rowbotham

Disruptive Technologies, Climate Change and Shipping
Edited by Barış Soyer and Andrew Tettenborn

DISRUPTIVE TECHNOLOGIES, CLIMATE CHANGE AND SHIPPING

EDITED BY

BARIŞ SOYER AND ANDREW TETTENBORN

CONSULTING EDITOR: SIMON BAUGHEN

informa law
from Routledge

First published 2022
by Informa Law from Routledge
4 Park Square, Milton Park, Abingdon, Oxon OX14 4RN

and by Routledge
605 Third Avenue, New York, NY 10158

Informa Law from Routledge is an imprint of the Taylor & Francis Group, an informa business

British Library Cataloguing-in-Publication Data
A catalogue record for this book is available from the British Library

Library of Congress Cataloging-in-Publication Data
Names: Soyer, Barış, editor. | Tettenborn, Andrew, editor.
Title: Disruptive technologies, climate change and shipping / edited by Barış Soyer and Andrew Tettenborn ; consulting editor: Simon Baughen.
Description: Abingdon, Oxon ; New York, NY : Routledge, 2022. | Series: Maritime and transport law library | Includes bibliographical references and index.
Identifiers: LCCN 2021041123 (print) | LCCN 2021041124 (ebook) | ISBN 9780367725358 (hardback) | ISBN 9780367725372 (paperback) | ISBN 9781003155195 (ebook)
Subjects: LCSH: Maritime law. | Technological innovations—Law and legislation. | Climatic changes—Law and legislation. | Maritime law—Environmental aspects.
Classification: LCC K1150 .D57 2022 (print) | LCC K1150 (ebook) | DDC 343.09/6—dc23/eng/20211105
LC record available at https://lccn.loc.gov/2021041123
LC ebook record available at https://lccn.loc.gov/2021041124

ISBN: 978-0-367-72535-8 (hbk)
ISBN: 978-0-367-72537-2 (pbk)
ISBN: 978-1-003-15519-5 (ebk)

DOI: 10.4324/9781003155195

Typeset in Times New Roman
by Apex CoVantage, LLC

CONTENTS

CONTRIBUTORS

EDITORS

Professor Barış Soyer
Professor of Commercial and Maritime Law
Director of the Institute of International Shipping and Trade Law
Swansea University

Professor Soyer directs the Institute of International Shipping and Trade Law at Swansea University, and is a member of the British Maritime Law Association and British Insurance Law Association. He is the author of *Warranties in Marine Insurance* (2001) (winner of the BILA Book prize in 2002), *Marine Insurance Fraud* (2014) (winner of the same prize in 2015) and has written numerous articles published in journals including the *Cambridge Law Journal*, the *Law Quarterly Review*, *Lloyd's Maritime & Commercial Law Quarterly*, the *Edinburgh Law Review* and the *Journal of Business Law*. He has also edited large numbers of collections of essays on commercial, maritime and insurance law. In addition, he sits on the editorial boards of the *Journal of International Maritime Law*, *Shipping and Trade Law* and the *Baltic Maritime Law Quarterly* and is on the editorial committee of the *Lloyd's Maritime and Commercial Law Quarterly* (International Maritime and Commercial Law Yearbook). He currently teaches Charter Parties and Carriage of Goods by Sea and Marine Insurance on the LLM Programme at Swansea and also is the Director of Commercial, Maritime and International Trade LLM Programmes.

Professor Andrew Tettenborn
Professor of Commercial Law
Institute of International Shipping and Trade Law (IISTL),
Swansea University

Professor Tettenborn has been attached to the IISTL at Swansea Law School since 2010, teaching international trade, aspects of banking, admiralty and energy law. He has also taught at the Universities of Cambridge, Exeter and Geneva, and held visiting positions in Europe, Australia and the USA. His book with Francis Rose on Admiralty Claims came out this year, and his extensively revised new edition of Marsden & Gault's *Collisions at Sea with John Kimbell* QC has just gone to press.

By way of background, Professor Tettenborn is also General Editor of the leading student textbook on commercial law (Sealy & Hooley's text, *Cases and Materials*) and Editor-in-Chief of *Clerk & Lindsell on Torts*. In addition, he has authored numerous articles on commercial law and obligations, and sits on the editorial boards of *Lloyd's Maritime &*

Commercial Law Quarterly and the *Journal of International Maritime Law*. As well as his other activities, he was also involved in the production of the English Restatements of restitution and contract law, and is also a member of the committee set up by Prof. Gerard McMeel QC in 2020 to produce a draft Commercial Code for England.

CONTRIBUTORS

Astrid Ainley
Trainee Solicitor
HFW Astrid

Ainley is a trainee solicitor at HFW. To date she has worked in HFW's commodities litigation, transactional insurance and the "Wet" shipping teams. During her time in shipping she has dealt with disputes involving salvage, collisions, vessel fires and groundings. She has worked with clients involved in proceedings in the English High Court and international arbitrations in London and Singapore. She is also a member of HFW's Autonomous Vessel Group. Prior to joining HFW she worked in the insurance industry specialising in long-tail claims and claims litigation.

Dr Zoumpoulia Amaxilati
Lecturer in Shipping and Trade Law
Institute of International Shipping and Trade Law,
Swansea University

Zoumpoulia Amaxilati is a Lecturer in the Institute of International Shipping and Trade Law (IISTL) at Swansea University, which she joined in 2019. She is a graduate of the Aristotle University of Thessaloniki and holds an LLM in Maritime Law from the University of Southampton, from where she subsequently obtained a PhD in 2019. Zoumpoulia teaches Admiralty law, Charterparties, Carriage of Goods by Air, Sea and Land, and also Tort Law.

Professor Simon Baughen
Professor of Commercial Law
Institute of International Shipping and Trade Law,
Swansea University

Professor Baughen was appointed Professor of Shipping Law in the Department of Shipping and Trade Law at Swansea University in September 2013. He had previously been a Reader at Bristol, where he had worked since joining academia in 1989. Simon studied law at Oxford and practised in maritime law from 1979 to 1988. His research interests lie mainly in the field of shipping law, in which he has written extensively over the last thirty years. He is the author of *Shipping Law*, now in its seventh edition, and in 2013 took over the editorship of *Summerskill on Laytime* in time for its fifth edition. The sixth edition was published in Autumn 2017.

He also researches in the field of the interaction between free trade/investment and environmental protection and human rights. He has produced two books on this topic. *International Trade and the Protection of the Environment* in 2007 and *Human Rights and Corporate Wrongs. Closing the Governance Gap?* in 2015.

Andrew Beale OBE
Visiting Associate Professor of Intellectual Property Law
Institute of International Shipping and Trade Law,
Swansea University

Andrew joined Swansea in 2002 as the Director of IP Wales, one of Swansea's award-winning business support initiatives. Originally based at the University of Wales Trinity Saint David in Swansea, Andrew became the Director of the Swansea Intellectual Property Rights Initiative in 1999. Andrew was responsible for designing and launching IP Wales in 2002. This organisation has been highly successful, gaining the Judges' Special Prize at the WORLD Leaders European Awards in 2004, and under Andrew's leadership has assisted over 800 Welsh businesses to make the most of their UK and international IP assets. In 2008 Andrew was seconded to work for WIPO, where he co-organised and presented at the WIPO Forum on Intellectual Property & SMEs for IP Offices of OECD and EU Enlargement Countries in Cardiff. Andrew remains as Director of IP Wales and was responsible for the validation of LLM programme in Intellectual Property and Commercial Practice at Swansea. Now in active retirement, his research interest lies in IP Cybersecurity and the online protection of trade secrets.

Andrew received the OBE for services to intellectual property and business in 2009.

Professor Olivier Cachard
Professor of International Commercial Law
Faculty of Nancy,
Université de Lorraine

Prof. Dr. Olivier Cachard is Full Professor (*agrégé*) of International Commercial Law at the Faculty of Nancy, Université de Lorraine, where he has established a LLM in International & European Business Law (now ranking 3rd best French program) and where he runs the double BA in Civil & Common Law. He has published eight books, including an *International Commercial Law Treatise* and an annotated *Maritime Code*. He is a member of the Academic Senate of the University of Lorraine, a member of the Board of the Association de droit maritime and a regular contributor to *Droit maritime français*. Olivier Cachard is acting as an Arbitrator, both *ad hoc* and under the rules of the *Chambre arbitrale maritime de Paris* and *ICC*. He also is an Appellate Lawyer and an elected Member of the Bar Council.

Dr. Michael Chatzipanagiotis
Lecturer in Private Law,
University of Cyprus

Michael Chatzipanagiotis is Lecturer in Private Law at the Department of Law in the University of Cyprus. His main research interests are AI, air and space law, insurance, intellectual property, banking and tort. He possesses a law degree from the University of Athens, Greece, and an LLM from Cologne, from which he also obtained a PhD *summa cum laude*.

He has written (with Dr. George Leloudas) on automated vehicles and third-party liability for an American journal and has contributed a chapter, *Technology & Policy*, to a forthcoming book (T. Synodinou et al. (eds), *EU Internet Law in the Digital Single Market*). He has also penned a chapter on *Insurance regulation and supervisory law of Cyprus* in P. Marano and K. Noussia (eds), *Transparency in Insurance Regulation and Supervisory Law*, and a conference paper on AI and liability under international space law, which was presented at the 71st International Astronautical Congress in 2020. Professionally Dr Chatzipanagiotis has worked as a legal adviser in Cyprus and Germany and as a lawyer in Athens, and participated as an air law expert in various EU programmes.

Henry Clack
Associate,
HFW

Henry Clack specialises in commercial dispute resolution, with a particular focus on both "wet" and "dry" shipping issues. At different times he has represented salvors, shipowners, P&I Clubs, commodity traders, brokers and logistics companies in disputes arising from matters ranging from salvage, fires and collisions to cargo contamination, charterparties and commodity supply contracts. He is also a member of HFW's Autonomous Vessel and Cyber Groups. Henry has represented clients in the High Court and in numerous international arbitrations under the LOF, LMAA and LCIA rules. During his training contract, he spent time in HFW's Geneva office where he worked on a broad range of commodity disputes (one of which reached the Supreme Court: *Taurus Petroleum v State Oil Marketing Co of Iraq* [2017] UKSC 64).

Julian Clark
Senior Partner,
Ince

Julian Clark is Global Senior Partner at Ince, responsible for its numerous sectors and vast client base across the globe. With over 30 years' experience in mediation, arbitration and litigation, Julian is an internationally recognised and influential leader in shipping and international trade, being consistently well ranked in ranked in Chambers & Partners, Legal 500 and Who's Who Legal.

Professionally he is a well-known expert in piracy, e-commerce and blockchain, cybercrime, security and terrorism. He is currently Chair of the Maritime London Innovation and Technical Committee, and of the Cyber Risk subcommittee, of the CMI.

Paul Dean
Global Head of Shipping,
HFW

Paul Dean is Global Head of Shipping at HFW, managing 200 lawyers across a worldwide network of 20 offices. He himself specialises in charters, carriage,of lading, salvage, towage, shipbuilding and offshore structures.

He is regularly seen at offshore supply vessel conferences and is an expert on SUPPLYTIME. He has taught on the BIMCO SUPPLYTIME panel for 10 years, and served on the review committee for the SUPPLYTIME 2005 revision and later on the committee set up to draft BIMCO's new standard form DISMANTLECON. In addition he contributed to the two most recent editions of Simon Rainey Q.C.'s *Law Of Tug and Tow and Offshore Contracts*. Not surprisingly he is very highly rated by the leading legal directories, and in 2019 and 2020 he was named by Lloyd's List as one of the top 10 maritime lawyers and one of the 100 most influential people in the maritime industry.

Professor Ellen J. Eftestøl
Professor of Civil and Commercial Law,
University of Helsinki and Adjunct Professor,
Scandinavian Institute of Maritime Law,
University of Oslo

Ellen is Professor of Civil and Commercial Law at the University of Helsinki. Her major teaching and research areas are related to international commercial transactions. Among other things she teaches contracts, torts, obligations and carriage. In addition, she is the founder and head of the interdisciplinary INTERTRAN Research Group for Sustainable Business and Law at the University of Helsinki. On occasion she acts as arbitrator in commercial law disputes.

Professor Eftestøl is joint author with Sankari and Bask of *Sustainable and Efficient Transport: Incentives for Promoting a Green Transport Market* and sole author of *Sustainable Carriage of Goods: The Role of Contract Law*.

Dr. George Leloudas
Associate Professor of Commercial Law
Institute of International Shipping and Trade Law,
Swansea University

Dr George Leloudas is Associate Professor at the Institute of International Shipping and Trade Law at Swansea, which he joined in 2011. He has an undergraduate degree from the Kapodistrian University of Athens, LLM degrees in Commercial Law from Bristol and in Air and Space Law from McGill, and a Cambridge PhD completed in 2009. At Swansea he teaches insurance, carriage, arbitration and aircraft finance.

George's principal research interest is the carriage by air, but his interests extend to multimodal transport, insurance and autonomous transport systems. He has published two monographs, one on *Risk and Liability in Air Law* and another on *Air Cargo Insurance* (the latter jointly with Professor Malcolm Clarke of Cambridge). He is also an editor of *Shawcross and Beaumont on Air Law*, being responsible for the passenger and cargo liability chapters. George has also written widely on carriage, fishing law and automated vehicles and vessels.

Dr Ewan McGaughey
Reader, King's College,
London At KCL

Dr McGaughey specialises in law, economics and history, with core research interests are economic and social rights, particularly in the governance of enterprises. He also doubles as Research Associate at the University of Cambridge's Centre for Business Research, teaches annually at the Paris School of Economics and is a volunteer for the Free Representation Unit. In the last few years he has held visiting positions at the universities of California at Berkeley, Fukuoka and Sydney.

David Owens
Managing Associate,
Ince

David Owens is a Managing Associate at Ince. Having previously worked as in-house counsel with a leading maritime operator, he is now a vital member of the Ince shipping team where he advises on issues of maritime law, litigation and arbitration.

John Russell QC
Quadrant Chambers

John Russell QC is a very experienced commercial advocate, both in the Commercial Court and above (he was the successful counsel in *Volcafe v CSAV*), and also in numerous international and maritime arbitrations. He was named Shipping Silk of the Year for the Legal 500 UK Awards 2020 and shortlisted for a similar title by Chambers & Partners in 2019. His expertise is wide, encompassing commercial dispute resolution generally with a particular focus on shipping, commodities, international trade, marine insurance, aviation and travel. He also acts regularly as an arbitrator.

Dr. Youri van Logchem
Senior Lecturer,
Institute of International Shipping and Trade Law,
Swansea University

Dr van Logchem is Senior Lecturer at the Institute of International Shipping and Trade Law at Swansea, where he teaches both undergraduates and postgraduates, primarily in the areas of the law of the sea, international energy law, and general public international law. Before that, he was a PhD fellow at the Netherlands Institute for the Law of the Sea at Utrecht. He has also been a lecturer and researcher at Utrecht, and a Junior Policy Officer at the Hague Institute for the Internationalisation of Law.

Youri's first degree was from Utrecht University, from which he also has a master's degree in legal research, and a PhD on the rights and obligations of states in disputed maritime areas.

He has published a good number of articles and contributions to books, some of which have been cited in international courts and tribunals. His monograph, *The Rights and Obligations of States in Disputed Maritime Areas*, is currently in press at Cambridge University Press. Youri has also received awards for his academic achievements, including the Rhodes Academy Submarine Cables Award, and has presented at various conferences around the world.

Emilie Yliheljo
Doctoral candidate, Faculty of Law and Institute of Sustainability Science (HELSUS), University of Helsinki

Emilie is a doctoral candidate at the Faculty of Law at Helsinki, writing a PhD thesis on the regulation of the European carbon market. She is a member of the INTERTRAN Research Group for Sustainable Law and Business. Emilie is specialised in climate law, climate finance and the governance of market-based instruments connected with climate policy. In addition to this, Emilie has over 10 years' professional experience working on emissions trading and climate finance in the public and private sectors, spending time (for example) as a specialist in the Energy Department of the Finnish Ministry of Economic Affairs and as a legal adviser on emissions trading at the Finnish Energy Authority. In 2015 she co-authored a report to the Nordic Council of Ministers on EU emissions trading scheme and its effect on road transport; more recently she contributed a book chapter on the prospects for expanding the EU emissions trading system to the transport sector.

PREFACE

This book has its origins in the 16th Annual Colloquium of the Institute of International Shipping and Trade Law held in September 2020 at Swansea University (online, owing to the COVID contagion). It deals with two significant contemporary issues now worrying shipping: the use of new and disruptive technologies, and the need to respond to climate change.

The shipping business is not for the faint-hearted: those in it have always been well-known for their entrepreneurial mindset. This essentially means that as long as there is money to be made from the exercise, they increasingly have no hesitation in employing new technologies, whether these be encryption, artificial intelligence, robotics, wireless sensing or full-fledged blockchain. Two legal issues immediately emerge out of this, however: the need for and scope of state regulation, and the extent to which private law liabilities emerging from the use of such technology can fit into the existing legal regimes.

The contributions in the first part of this volume deal with these fundamental issues. We have, for example, novel and penetrating debates on the use of distributed ledger technology in various aspects of shipping and insurance context; the prospect of utilising model laws to support the development of e-shipping; the deployment of drones in maritime context; various aspects of autonomous shipping; and the potential impact of cyber risks in this context. And let us not forget something which is often forgotten: IP-related issues emerging from disruptive technologies employed in shipping.

The second part of the book is devoted to shipping and climate change. Hitherto shipping, never an industry voluntarily to adopt costly measures unless compelled to, has been one of the slowest businesses to take serious steps to protect the environment. But no longer: the law is fast catching up, as is environmental consciousness among customers. Hence the coverage here. We have concentrated, from the point of view of English, EU and international law, on the specific effects of climate change and associated environmental risks on the shipping sector. A number of highly eminent contributors have dissected key issues such as international law measures to reduce carbon emissions in shipping; the rights of various parties in environmentally sensitive areas such as the Arctic; maritime emission trading systems; civil liability for provoking climate damage; and the long-term aim of complete elimination of fossil fuels.

We are enormously grateful to many who behind the scenes made possible the Colloquium and this book that grew from it. Our Research Assistant, Alicia McKenzie, was indispensable and unflappable in providing the essential unsung back-up without which these events just cannot happen. Our publishers, Informa Law, again provided us with their unstinting support as they had with the previous events. Lyn Reynolds, our events co-ordinator at Swansea, directed our first digital Colloquium run with military efficiency and great humour. We are also grateful to the contributors and those who participated at our first online Colloquium for their insightful comments, and to our audience for coming up with difficult and insightful questions. Those certainly helped all of us in polishing up our contributions.

The law in this area is fast developing. We sincerely hope that the book will substantially contribute to the learning and understanding of it.

Professors B. Soyer & A. Tettenborn
Swansea, June 2021

TABLE OF CASES

TABLE OF LEGISLATION

Statutes

Statutory instruments

European Legislation

International Legislation

CHAPTER 1

Shipping and distributed ledgers

Of paper, code and progress

*Andrew Tettenborn**

1.1 Introduction

Paper fills filing cabinets and empties forests. Because of its mutability, destructibility and propensity to get mislaid, it also on occasion gives lawyers nightmares. Electronics, by contrast, merely consume current. This chapter discusses how far paper can and should give way to electronics in the context of shipping law.

In the field of commercial law, the use of paper raises four classic difficulties. One relates to the authentication of documents, and in particular how you can effectively prevent people creating duplicates or forgeries. The second problem is that of safe storage, and the making available of what is in documents to those (and only those) with a need or right to see it. The third is the prevention of later unauthorised tampering. The final one is difficulties in quick and secure transmission. In the nature of things, paper can travel only as fast as the swiftest courier or Boeing 787, and is in addition liable to physical loss or mutilation en route.

Digitisation has been with us for some time. But although it deals with the last of these problems (as for that matter does fax), it does little about the others. Computer storage and email are of course useful in their way, but both have always been relatively hackable and on their own provide little serious security. Again, a measure of copy-protection of electronic documents is undoubtedly available, but is still fairly rudimentary in the face of a determined fraudster. Meanwhile the universal availability of sophisticated printing software has greatly eased the forger's task; a bogus but plausible bill of lading can now be produced in hours, if not minutes. Furthermore, even to the extent that it is more difficult to interfere with documents, the time it takes visually to inspect them and detect forgeries sits increasingly ill with carriers' and shippers' ever-augmenting desire for speed of turnaround above everything else.

In many cases this actually does not matter, since the problems outlined above are not very significant anyway. Where this is so, the limitations of digitisation are likely to be no big deal. For example, contracts of affreightment and charterparties, or supply and bunkering contracts, have in practice for some time been largely negotiated and concluded electronically by exchange of emails or some similar process: no paper vs electronics issue arises here, since such a contract can be made in any form and the presence or absence of a paper copy of it is a legal irrelevance. Slightly less obviously, many of the same arguments can be applied to contracts of carriage where there is no question of any need for negotiable carriage documents. It is a matter of some indifference whether, say, a sea waybill or ferry

* **Professor of Commercial Law in the Institute of International Shipping and Trade Law, Swansea University.**

DOI: 10.4324/9781003155195-1

consignment note is concluded in written or electronic form: this is largely a matter of the parties' convenience, since its validity and legal effect are the same in either case.[1]

Outside these cases, however, something more is needed. It is here that we see the potential relevance of the new development known as distributed ledger technology, and its offshoot blockchain.[2] These techniques have the potential to make the production and use of forgeries difficult, the interception or unlawful copying of a document in transmission vastly harder and unlawful tampering with an existing electronic document almost impossible. This is particularly relevant in three areas: shipping documentation, cargo care and vessel monitoring. In this chapter will deal with the second and third issues.

1.2 Cargo, damage and monitoring

Unlike the application of blockchain technology to carriage contracts such as bills of lading, which raises serious technical legal issues (such as whether paper documents can be replaced by electronic impulses and still retain validity as a matter of law, or what counts as delivery or possession of a disembodied stream of electrons), with cargo care the matter is simpler. Cargo litigation, for all its complexity, is overwhelmingly fact-specific. What happened to the cargo, when and why? The relevance of documents, electronic or otherwise, is not so much legal as evidentiary. This relevance is twofold. First, the carriage process is, in practice, very extensively documented,[3] which means that in more and more cases the outcome will depend on what these documents say. Second, carriage is subject to a number of legal presumptions and other provisions (for example, the rule that a bailee is presumptively liable for damage or loss unless he proves an excepted peril),[4] with the shipping documents being strong evidence as to state and quantity of goods originally entrusted to the carrier.

So too in the converse case of reverse cargo litigation: where damage is allegedly caused by cargo to a vessel or other cargo on board during a voyage, what we need in order to know whether a dangerous cargo claim is feasible is when and how the damage happened. Once again, this is simply a matter of evidence; once again, in practice the documentation will be crucial.

Distributed ledger technology has the potential to be highly significant in this respect, especially when combined with the fact that continuous surveillance of cargo en route is

1 It might be said that ss. 2 and 3 of the Carriage of Goods by Sea Act 1992, referring to the rights and duties of the holder of a sea waybill, envisage possession of a physical piece of paper. But they say nothing about any need for physical transfer or delivery of a relevant document. It seems to follow that that if a person without a physical document in his hands wishes to constitute himself such a holder, he can do this by the simple expedient of printing out an electronic document at the nearest Kall-Kwik or indeed from his own HDD.

2 "Blockchain" and "distributed ledger" are often used as if they were synonyms. But technically the former is a subgroup of the latter. "Distributed ledger" is any system in which a database is distributed over a number of participants, rather than being dependent on a centralised server, with changes having to be made to all copies. "Blockchain" is a distributed ledger system where each ledger is readable by all participants and all ledgers can be securely and indelibly updated virtually simultaneously, with all additions and their timing being permanently recorded. See I. Bashir, *Mastering Blockchain: Distributed Ledger Technology, Decentralization, and Smart Contracts Explained* (2nd edn, 2018, Packt Publishing), Chapter 1.

3 Documentation increasingly required by law: see e.g. the introduction in July 2016 of the requirement to ascertain a VGM for all containers prior to loading: SOLAS, Ch.VI, Regulation 2.

4 Most recently elucidated in *Volcafe Ltd v Compañía Sud Americana de Vapores SA* [2018] UKSC 61; [2019] A.C. 358.

rapidly becoming the new normal.[5] This is particularly true with containers, which are progressively becoming not simply sealed boxes, but smart machines equipped with sensors.[6] The latter are able to detect the temperature of a freezer or reefer container and any changes in it, watch the humidity in a dry container or detect sudden movement in any box. They can equally check for weight,[7] listen for footsteps, detect door openings and film interior events and where necessary provide a precise geolocation. In the longer term it may be possible for them to detect the nature of a given cargo. The practical applications go in step.[8] Precise geolocation means shippers and consignees can be provided with real-time information as to where their goods are and when they are likely to get them. Sudden changes in temperature or atmosphere can be fairly precisely timed, thus implicating the carrier if the goods should arrive damaged or deteriorated (or for that matter exonerating it in the case of through or multimodal carriage, in so far as it indicates that the problems actually occurred at some other stage). Sensors can equally be programmed to avoid problems in the first place by triggering compensating adjustments – for example, changing the ambient atmosphere in a container full of fruit, or dehumidifying a cargo of coffee to ensure its arrival in perfect condition – or in extreme cases by initiating such things as fire precautions. Theft and pilferage from a container become more difficult when door opening is constantly monitored and there are cameras inside; fraudulent removal of the container itself (for instance, in the course of discharge) becomes much more problematical if those doing it know that geolocation may be active. As and when accurate identification of cargo in a container becomes possible, carriers will be able more effectively to deal with the problem of undeclared dangerous cargo,[9] and so on.

Such developments are most clear with containers, but it should not be forgotten that bulk cargoes such as gas, chemicals, oil or grain are equally subject to increasing automatic monitoring by equipment on board,[10] and most of the comments above on containerised cargo will apply at least to some extent to them too.

On its own, an onboard electronic information gathering system largely benefits shipowners (or bareboat charterers) acting as physical carriers; they are, after all, in control of the vessel, they will receive the information and act on it[11] and they will ultimately be

5 Not only at sea, either: this reflects a more general trend. See "Logistics 4.0: How IoT Is Transforming the Supply Chain" (*Forbes*, 14 June 2018), describing the tendency towards continuous monitoring from manufacture to end-use, using blockchain to ensure the data emanating from it is reliable and tamper-proof. The article is available at www.forbes.com/sites/insights-inteliot/2018/06/14/logistics-4-0-how-iot-is-transforming-the-supply-chain/#7eea5b19880f (last accessed on 31 January 2021).

6 See e.g. F. Stevens, "Smart Containers: The Smarter, the More Scope for Liability?" in Chapter 5 of B. Soyer and A. Tettenborn (eds), *Maritime Liabilities in a Global and Regional Context* (Informa Law, 2018); and for a more recent coverage: "The Power of Parameters in Smart Container Solutions" in *Maritime Executive*, 28 April 2020.

7 Important since the introduction in July 2016 of the requirement to ascertain a VGM for all containers prior to loading: SOLAS, Ch.VI, Regulation 2.

8 They are nicely summed up in a 2019 White Paper issued by the UNECE (UN Commission for Europe), "Smart Containers: Real-time Smart Container data for supply chain excellence" (ECE/TRADE/C/CEFACT/2019/10) p. 9 ff.

9 On this, see "Can Blockchain Solve the Problem of Misdeclared Dangerous Goods?" in Marine Log for 26 June 2019 (available on line at www.marinelog.com/news/can-blockchain-solve-the-problem-of-misdeclared-dangerous-goods/, last accessed on 31 January 2021).

10 See for example, the IMO's IGC, IBC and IMSBC Codes.

11 One advantage seen by carriers as stemming from information about the contents of containers is valuable marketing intelligence as to their customers' wants and needs.

the party responsible for ensuring that any necessary measures are taken.[12] But there is no reason why it should be made available to them exclusively: distributed ledger technology makes it possible for it be distributed to others involved in the operation.[13]

Of course, whether information is shared in a given situation will ultimately be governed by commercial considerations. Nevertheless, it may well be to the benefit of all parties for reliable and unmanipulable cargo data to be shared between two or more of, say, carriers, shippers, consignees and container suppliers (and possibly their respective insurers or P&I Clubs). The point is that in many cases they will share a common interest in receiving reliable information. A great deal of the expense of cargo disputes, for example, result from arguments over when damage occurred – especially in the case of serial carriers, a point arising not only in connection with transhipped cargo but also in almost every case of multimodal transport – and what its immediate cause was. A sharing of reliable cargo monitoring data between the carrier (or carriers) and the shipper will normally avert argument over the former and will often help with the latter. Similarly with arguments about whether damage occurred before or after shipment: in so far as a smart container will have the capacity to begin monitoring at the time of stuffing, the prospect of being able to establish whether damage occurred after that moment, and if so when, is attractive. And so also with dangerous cargo claims, especially if and when reliable technology becomes available to identify otherwise invisible hazardous cargo, as well to monitor such matters as combustion.

1.3 Vessel monitoring

Apart from issues connected with cargo, the decreasing size of crews and the increasing complexity of the vessels they man mean that electronics are now indispensable for the monitoring of the well-being and progress of any cargo vessel. Much of this electronic input, indeed, is now required by law. For some twenty years all vessels of any size[14] have been bound under SOLAS to carry electronic VDR equipment;[15] for the last sixteen the same has applied to AIS systems capable of transmitting details to interested parties of a vessel's identity, type, position, course, speed and navigational status.[16] In addition, there are numerous other requirements and recommendations applicable in particular trades or to particular vessels: to take just one example, sensitive hull stress monitoring machinery on bulk carriers.[17] Quite apart from any legal requirements, however, the commercial self-interest of owners and operators desperate to cut crewing costs means that the monitoring of the inner workings of a vessel is now very much taken care of by of automated processes.

This has already made a clear impact in civil litigation. For instance, in an Admiralty collision case all VDR data must now be exchanged at an early stage under the CPR;[18] and

12 How readily such information is made available to cargo interests varies. The writer understands informally, for instance, that Maersk and Hapag-Lloyd fairly readily hand it over, whereas other carriers are more secretive.

13 A point made in "Standards for Smart Container Data Exchange Published", *Maritime Executive*, 6 October 2019.

14 That is, over 3,000 GRT.

15 See SOLAS, Ch.V, Reg.20.

16 For details see SOLAS, Ch.V, Reg.19.

17 See the IMO recommendation contained in MSC/Circ.646 (6 June 1994).

18 See CPR 61.4(4A).

today it is by no means impossible for such suits to be decided on the basis of little else.[19] There is obviously scope for extending this further, for example to grounding cases.[20]

More importantly, however, there is obvious scope for blockchaining data of this sort to preserve its existence and integrity. Partly this is a matter of public law, on the basis that the preservation of information in the hands of shipowners, registry states tasked with port state control and others can be vital for full and proper accident investigation (though it must be admitted that despite the existence of facilities to do this[21] take-up has not been impressive).[22] But even putting this aside, there may be considerable scope for sharing this information in the context of private law relationships, just as we suggested there was in the case of cargo. Two areas seem particularly promising here. One is charters; the other, insurance.

1.4 Charter disputes

Distributed ledger technology is beginning to affect the charter market, especially when it comes to such logistical matters as the arranging of fixtures and the juggling of available dates.[23] But there ought to be scope for it to go a great deal further. A lot of charter disputes, especially those arising under time charters (but not forgetting also a few arising under bareboat charters), concern matters where shared trustworthy information is at a premium. Think, for example, about questions of what caused, and hence who is responsible for paying the costs of, accidental damage to a vessel in the course of a voyage; or whether a cargo or demurrage claim can be passed up (or sometimes down) a line of charterers and subcharterers. Not that these are the only instances: other examples include whether a vessel has infringed the geographical, or possibly the cargo, limits laid down in the charter, and if so, when; whether a duty to proceed with dispatch, or a speed or consumption warranty, has been broken; and so on. And this of course is in addition to the usual disputes about whether the necessary criteria are satisfied to put a vessel off-hire under whichever clause applies in the individual case.

The legal answers to these questions will admittedly vary greatly. They will always turn on the form of charter used, and any amendments to it; as any charter lawyer will confirm, a NYPE 2015 charter may very well give different legal answers from a Beepeetime 3, the latter being a markedly more pro-charterer document. But in almost every case the issue will turn on points of fact: where the vessel was at a particular time, what was being carried, the vessel's progress at a point in time, events taking place on board her at a particular

19 For an example of this process in action, see the recent decision of the Admiralty Judge in *Nautical Challenge Ltd v Evergreen Marine (UK) Ltd (No. 2)* [2017] EWHC 453 (Admlty); [2017] 1 Lloyd's Rep. 666.

20 See e.g. the grounding case of *Alize 1954 v Allianz Elementar Versicherungs AG* [2019] EWHC 481 (Admlty); [2019] 1 Lloyd's Rep. 595.

21 For example, through Inmarsat's Fleet Data System and offerings from other private enterprise operators. See "Inmarsat Unveils Major New IoT Service for the Shipping Industry" (News release, 4 September 2018, available at www.inmarsat.com/news/inmarsat-unveils-major-new-iot-service-for-the-shipping-industry/, last accessed September 2020).

22 No doubt because, in the case of casualties, shipowners and others wish to minimise the possibility that matters possibly demonstrating their own culpability will reach either the public via news or social media, or the relevant prosecuting authorities, to whom it might give ideas.

23 The Antwerp-based Shipnext platform (www.shipnext.com), for example, includes among its services what is essentially a blockchain-based shipbroking platform for matching not only all sorts of ships and cargoes, but also, further up the chain, disponent owners and potential time charterers.

moment, and so on. And all these are points on which there will be automatic electronic records, in nearly all cases today, whether required by law[24] or merely a matter of practice.[25]

It is true that if such matters come to litigation, then the process of disclosure – which in England is normally fairly reliable, if time-consuming and laborious – will go a long way towards dealing with the problem of establishing the true facts on which the answer to the charter dispute turns. But ideally such disputes should be smartly settled, rather than litigated lengthily; and in this case there is every interest in speeding up the process as far as possible. There is much to be said for enabling the necessary information to be shared in reliable and unalterable form in real time between those concerned – notably, owners, charterers and subcharterers and their respective insurers and P&I Clubs – under a distributed ledger agreement, so that if a dispute does arise the facts can be established as fast as possible. Admittedly the shared information would have to be selected carefully, for reasons of commercial confidentiality, and access to it equally carefully limited to those with a genuine interest in having it. But that is exactly the strong point of distributed ledger technology; its sophistication today would seem to make this entirely possible. It remains to be seen whether the possible savings in dispute resolution costs will persuade shipowners to ditch their natural reluctance to share information with others in the transport logistics chain. The matter remains uncertain; nevertheless, there are good reasons to hope that they will.[26]

1.5 Insurance

As with chartering, so with insurance. The industry has recognised for some time the need to move beyond word-processor and paper as regards the setting up of insurance arrangements, the estimation of risk and the setting of premiums. All involve large information, organisation and processing costs, and these costs need streamlining in order to maintain competitiveness. In this area, it is not surprising that experiments with shared information using distributed ledger or blockchain technology are well advanced;[27] and indeed, in addition there is a possibility of using some kind of smart contracts technology to pay at least run-of-the-mill claims.[28] In another branch of the insurance industry, P&I clubs have also engaged with distributed ledgers, but mainly in connection with their coverage of cargo liabilities, where they have progressively approved various forms of blockchain-based systems such as BOLERO and essDocs for replacing paper shipping documents with electronic ones.[29]

---

24 For example, the VDRs mandated by the provisions of SOLAS V, Reg.20, mentioned above.

25 For example, cargo and fire monitoring equipment over and above that required in such documents as SOLAS II-2, Reg.7.

26 A consortium has already been set up by the Lloyds Register Foundation and others to apply blockchain to the solution of dangerous cargo problems (see www.marinelog.com/news/can-blockchain-solve-the-problem-of-misdeclared-dangerous-goods/, last accessed 31 January 2021).

27 For example, Ernst & Young and Guardtime have set up a digital pathway known as Insurwave, an experimental blockchain platform in collaboration with Maersk in order to streamline, among other things, the setting of premiums in hull insurance. The scheme is described in some little detail at www.ey.com/en_uk/blockchain/how-blockchain-is-reducing-fluidity-of-risk-in-marine-insurance (last accessed on 31 January 2021).

28 See generally, A. Borselli, "Smart Contracts in Insurance: A Law and Futurology Perspective" in P. Marano and K. Noussia (eds), *Insurtech: A Legal and Regulatory View* (Springer, 2019) 101–125.

29 See most recently the International Group's circular of 12 June 2019, "International Group Circular – Bills of Lading and Blockchain Based System" (available at www.igpandi.org/article/international-group-circular-bills-lading-and-blockchain-based-system, last accessed September 2020).

But there is, one would have thought, scope to go further. If the average charter dispute is heavily fact-specific and often dependent on matters covered by one or more kinds of electronic recording or surveillance equipment, the same applies in spades to the process of deciding whether an insurance policy will respond to a claim, and resolving the matter in the event of arguments. It is in the nature of insurance that precise information as to the surrounding circumstances is vital, both in deciding whether the facts put the claim is within the cover, and if so, whether it is nevertheless excluded by some exception such as unseaworthiness, or by some misrepresentation or non-disclosure by the assured. Engaging in highly time-consuming efforts to establish the true facts in such cases only adds to costs and inefficiencies.

If this is right, there is a good deal to be said for insurers and assureds setting up a shared-platform blockchain arrangement under which information on the workings of the vessel is shared between them, so that in the event of a claim arising the true situation can be settled as soon as possible, and payment made or withheld accordingly. A number of operations have been set up to do precisely this in cargo insurance, by making available cargo sensing devices and using their data to establish what has happened;[30] there is every reason to think that hull insurers have every interest in following suit.[31]

1.6 Conclusion

There is little doubt that blockchain, with its ability to tidy and streamline the production and use of documents, will affect shipping in the next few years in almost as many ways as it affects other businesses. This chapter has done little more than scratch the surface of its potential application in shipping law. It is largely speculative, if only because the blockchain itself is still in its infancy and shipping lawyers and others are still feeling their way as regards how to use it.[32] Nevertheless, it seems a racing certainty that the next ten years will see practice transformed in at least some of the areas mentioned. For once we might even be able to say that, at least to some extent, the shipping industry is at the forefront of technological development. It will make a nice change from castigating it, as we have frequently and habitually done in the past, for its innate stick-in-the-mud conservatism.

30 The Ascot and Beazley insurance groups, for example, announced in early 2019 that they had set up such a platform: see the report in Insurance Business for 13 February 2019 (available at www.insurancebusinessmag.com/uk/news/marine/lloydsbased-cargo-consortium-sets-sail-158716.aspx, last accessed September 2020).

31 For a comprehensive analysis on the potential use of blockchain applications in insurance sector, see Chapter 3.

32 And for that matter the rules of the general law are only beginning to be methodically updated to take account of it. It was only in the second half of 2020 that the Law Commission finally started its general project on "adapting English law for the digital revolution".

CHAPTER 2

Blockchain and electronic bills of lading

Can revolutionary technology facilitate evolutionary change?

*John Russell**

2.1 Introduction

This contribution considers whether the use of blockchain might (finally) facilitate a move to a wider adoption of electronic bills of lading as a replacement for paper Bills.[1] It will first address the historical development and the modern functions of the paper Bill of Lading. It will consider some of the problems of the paper Bill, but also the problems that have been encountered with any widespread shift to electronic bills. Finally, it will discuss how blockchain may offer a way round at least some of these problems, and how this revolutionary technology could, perhaps, facilitate an evolutionary development in law and practice.

2.2 The Bill of Lading: the historical context

The paper Bill of Lading is the foundation stone of the English law of the carriage of goods. That remains as true today as it ever has done. And in this context, that is almost forever. In exploring the origins of the Bill of Lading Sir Richard Aikens et al.[2] observe: "For the purposes of our consideration, it is safe to say that in the eleventh century the bill of lading was probably unknown." The earliest Bill of Lading for which there is good evidence dates from the late fourteenth century.[3] It reads:

> 1390, the 25th day of June. Know all men that Anthony Ghileta shipped certain wax and certain hides in the name and on behalf of Symon Marabottus which things must be delivered at Pisa to Mr Percival de Guisulfis, and by order of the said Mr Percival who shall deliver all his things to Marcellino de Nigro his agent, and I Bartholomeus de Octono shall deliver all his goods at Portovenere and for the better caution I affix my mark so.
> A copy
> Bartholomeus de Octono mate of the ship of Anrea Garoll.

The similarities between the wording of the fourteenth century Bill, and of those in use today are striking. That said, it appears that in the early days a Bill of Lading served principally, and perhaps solely, as a receipt for the good shipped. Changes in trading practices, however, drove changes in the functions of the Bill of Lading. First, by the sixteenth

* **Barrister, Quadrant Chambers**

1 The chapter will use capitals where referring to paper Bills and not capitalise reference to electronic bills, since, for the moment at least, they are not Bills of Lading, properly so called.

2 *Bills of Lading* (3rd ed), (Informa, 2021), at 1.

3 E. Bensa, *The Early History of Bills of Lading* (Caimo & C, 1925), at 8.

 DOI: 10.4324/9781003155195-2

century, merchants had abandoned their earlier practice of generally travelling with their goods on the ship that they had chartered. Further, it was increasingly common for cargoes to be despatched before the shipper knew who the final receiver at the discharge port would be. Increasingly, the Bill of Lading was used as a means of transferring title in the goods, and finally, as a means of giving the holder of the Bill of Lading symbolic possession of the goods themselves. Thus, in *Lickbarrow v Mason*, confirming that (at least when the Bill was transferred pursuant to an underlying transaction) the transfer of a Bill of Lading also transferred property in the goods, Buller J (in the original King's Bench decision) held that:

> it appears that for upwards of 100 years past it has been the universal doctrine at Westminster-Hall, that by a bill of lading, and by the assignment of it, the legal property does pass . . . If these cases be law, and if the legal property vested in the plaintiffs, that, as it seems to me, puts a total end to the present case; for then it will be incumbent on the defendants to show that they have superior equity which bears down the letter of the law, and which entitles them to retain the goods against the legal right of the plaintiffs.[4]

Then, in the seminal decision of *Barber v Meyerstein*, the law took the important step of recognising that the transfer of the Bill of Lading gave its new holder symbolic possession of the goods to which it related. In the House of Lords, Lord Hatherley noted the principle:

> There has been adopted, for the convenience of mankind, a mode of dealing with property the possession of which cannot be immediately delivered, namely, that of dealing with the symbols of the property. In the case of goods which are at sea being transmitted from one country to another, you cannot deliver actual possession of them, therefore the bill of lading is considered to be a symbol of the goods, and its delivery to be a delivery of them.[5]

Lord Hatherley adopted a passage of Willes J in the court below, which uses language still much in use today:

> During this period, therefore, ***the bill of lading would not only***, according to the usage, and for the satisfaction of the wharfinger that he was delivering to the right person, ***be a symbol of possession, and practically the key of the warehouse***, but it would, so far at least as the shipowner was concerned, retain its full and complete operation as a bill of lading, there having been no complete delivery of possession of the goods.[6]

A further development in commercial practice was that, increasingly, a large number of cargoes, shipped by different merchants, came to be carried on each ship. This meant that the entering of individual charterparties for each parcel of goods became increasingly impractical, and there was a move, albeit rather stuttering, towards recording contractual terms in the Bill of Lading itself. The development of the Bill of Lading as a contractual document, or at least, as a means of transferring contractual rights and obligations, ran into the English common law's adherence to, or perhaps obsession with, the doctrine of privity. Thus, in *Howard v Shepherd* a reliance on the custom of merchants to support a claim that

4 (1787) 2 T.R. 63, at 69 (original King's Bench decision); (1790) 1 H. Bl. 357 (Exchequer Chamber); (1793) 4 Brown 57; (1793) 5 T.R. 367; (1793) 2 H. Black. 211 (House of Lords); (1794) 5 T.R. 683 (*venire de novo*) and (1794) 6 T.R. 131 (costs).

5 (1870) L.R. 4 H.L. 317, at 330.

6 Ibid., at 332 (emphasis added).

there was a contractual obligation on the carrier to deliver to the holder of the Bill of Lading was dismissed. Maule J held:

> If, however, the declaration is in case, it proceeds upon a supposed liability arising out of a contract transferred by the indorsement of certain bills of lading. Now, it is perfectly clear that a contract cannot be transferred so as to enable a transferee to sue upon it.[7]

The refusal by the courts to facilitate a bypassing of the doctrine of precedent led to statutory intervention, with the Bills of Lading Act being passed in 1855. The purpose of the 1855 Act, in relation to the transfer of rights, was set out in clear terms in the preamble:

> WHEREAS by the Custom of Merchants a Bill of Lading of Goods being transferable by Endorsement the Property in the Goods may thereby pass to the Endorsee, but nevertheless all Rights in respect of the Contract contained in the Bill of Lading continue in the original Shipper or Owner, and it is expedient that such Rights should pass with the Property.

Unfortunately, numerous difficulties with the implementation of this clear purpose arose in the cases that followed, leading, ultimately, to the Carriage of Goods by Sea Act 1992 ("COGSA 1992"), which has resolved most of the difficulties, if not all of them. Why the potted history? It is submitted that it illustrates the preference of the shipping industry for evolutionary change in relation to the legal frameworks which it adopts. The Bill of Lading, a document originating in the fourteenth century, has evolved far beyond its genesis as a mere receipt for goods.

2.3 The modern functions of the bill of lading

The historical development discussed here leads to the modern position where it is frequently said that a Bill of Lading has a tripartite function. Thus, in the Law Commission's report that preceded COGSA 1992, the Commission said:[8]

> We have also opted against a definition of "bill of lading", just as there is no definition under the 1855 Act or the Factors Acts. Under the present law, a bill of lading is usually identified by reference to its three functions, i.e., that it is a receipt for the goods, that it usually evidences the contract of carriage and that it may be a document of title (at least until complete delivery of the goods has been made to the person entitled thereto).

In this context, "document of title" does not necessarily connote that the holder owns the goods: very often the holder will not be the owner, for example a bank holding a Bill pursuant to a pledge. Rather the holder has the right to possession of the goods, and to have them delivered to him – it is the "key to the warehouse". Further, of course, not every Bill will perform all three functions, or not at all times. A clear example is that the Bill of Lading, in the hands of a charterer, does not contain or evidence the contract of carriage between the shipowner and the charterer (however much on its face it might purport to do so).

It is perhaps helpful to unpack the second function identified by the Law Commission, the contractual function, further. Not only does the Bill of Lading record or incorporate

7 (1850) 9 C.B. 297, at 319.

8 The Law Commission and Scottish Law Commission, *Rights of Suit in respect of Carriage of Goods by Sea* (1991, Law Com No. 196, Scot Law Com No. 130), at 2.50.

the contractual terms agreed between the shipper and carrier, as any contractual document might, rather, in addition:

i) It is one of the principal vehicles by which the contractual rights and obligations of the shipper may be transferred to a subsequent holder of the Bill, pursuant to COGSA 1992, in defiance of the common law rules of privity.
ii) It is one of the principal vehicles by which the Hague-Visby Rules may have the force of law in relation to a contract of carriage, pursuant to the Carriage of Goods by Sea Act 1971 ("COGSA 1971").

It is the function of the Bill of Lading as a vehicle for the transfer of contractual rights and obligations that has, at least arguably, been one of the biggest barriers to the widespread adoption of electronic bills of lading, as will be discussed below.

2.4 The problems with paper Bills

There are numerous problems with the existing paper-based system. First, there are practical problems. There is very often a disparity now between the speed at which the goods covered by a Bill of Lading physically move around the world, and the speed at which original Bills of Lading crawl through the trading and banking systems. Very often delivery of the goods will not be against presentation of the Bill, but against a Letter of Indemnity ("LOI") provided to the carrier. There is a disconnect between the theory of the Bill of Lading being the "key to the warehouse" and the way goods are often delivered in practice. It was striking, in a recent London arbitration, how the shipowners viewed Bills of Lading and LOIs as effectively interchangeable when it came to the delivery of goods. The system is supposed to be based, as a matter of theory, on the Bill of Lading as the symbolic key. As a matter of practice, it is probably now more often the LOI that opens the door. These problems have been significantly exacerbated by the COVID-19 pandemic which has made the physical endorsement and transfer of Bills of Lading all the harder given lockdowns, office closures and flight restrictions. In 2020 it was not uncommon for Bills of Lading to be waiting endorsement or transfer many months after the goods had been discharged from the ship and delivered to the receiver.

Second, the production of the paper, and the transfer of the paper, inflates transaction costs. In 2014 Maersk calculated that the documentation for a single shipment of refrigerated goods from East Africa to Europe could pass through up to 30 people and organisations, involving more than 200 different interactions and communications.[9] In addition to the human resources costs, there are the costs of printing, couriering and so forth. Hard figures are hard to come by, but some estimate that there are as many as 400 million shipping documents in circulation annually, and the total costs of document processing in the paper-based system may be as much as US$420 million a year. For so long as the physical piece of paper is the "key to the warehouse" or the physical transfer of the piece of paper is necessary to effect the transfer of contractual obligations, these costs are locked into the system. Again, these costs have been exacerbated by the pandemic.

9 Cited here at www.maritime-executive.com/article/zim-trials-blockchain-bill-of-lading and here https://nortonrosefulbright.com/en/knowledge/publications/b20094b6/e-bills-of-lading#section6 (accessed 1 February 2021).

Third, the number of human interactions increases the risk of human error. And, due to the importance of the original documents, the correction of errors, even once detected, is far from straightforward.

Fourth, again the focus on the physical transmission of original documents is a barrier to easy trading, for example in relation to the issue of new Bills where a trader wishes to split a cargo covered by a single Bill into smaller parcels. What should be a simple and low-cost process is much more labour intensive and costly than it could (and should) be.

Fifth, there is the risk of fraud. Despite the dependence on the physical transfer of original Bills, vanishingly few Bills contain any security features at all. Even when paper documents do contain security features, sophisticated fraudsters are able to replicate them (as the arms race between bank note producers and forgers attests) or they may not be used properly in practice. For example, in *Natixis v Marex; Marex v Access World Logistics*,[10] a high value nickel fraud case involving forged warehouse receipts (which share some features with Bills of Lading), there were some features of the forgeries which Bryan J described as being "unsophisticated and a dead give-away" but nonetheless the forgeries were passed as authentic by the warehouse staff employed to check them. Further, some more advanced security features in the receipts were not checked at all.

Sixth, and more recently, there have been environmental concerns raised about, simply, the production of all that paper, and perhaps more pertinently, the carbon and other emissions generated by the couriering of the paper (often by air) around the globe. As the shipping industry as a whole moves towards a more environmentally neutral model, it does seem bizarre to have air courier services embedded as an intrinsic part of it.

Given all these problems with a system dependent on insecure pieces of paper sluggishly but expensively moving through the mercantile and banking chains, it might have been thought that there would have been a rapid adoption of technology-based solutions. That has not proven to be the case, so far.

2.5 A short history of electronic bills

Electronic bills of lading are not a new concept. As long ago as 1990 the Comité Maritime International ("CMI") published its *Rules for Electronic Bills of Lading*.[11] COGSA 1992 recognised the possibility of electronic bills of lading fitting into the existing legal regime. As will be set out below, this may be critical to the facilitation of the widespread adoption of electronic bills based on blockchain technology.

Section 1(5) of COGSA 1992 as amended provides as follows:

> (5) The Secretary of State may by regulations make provision for the application of this Act to cases where an electronic communications network or any other information technology is used for effecting transactions corresponding to –
> (a) the issue of a document to which this Act applies;
> (b) the indorsement, delivery or other transfer of such a document; or
> (c) the doing of anything else in relation to such a document.

10 [2019] EWHC 2549 (Comm); [2019] 2 Lloyd's Rep 431.

11 This can be found at: https://comitemaritime.org/work/rules-for-electronic-billing-of-lading/ (accessed 1 February 2021).

We will return to this important section later. The year 1999 saw the introduction of the BOLERO bill of lading[12] and the publication of the first edition of the BOLERO Rule Book,[13] one of the pioneering electronic bills of lading systems. In December 2008 the UN General Assembly adopted the Convention on Contracts for the International Carrying of Goods Wholly or Partly by Sea. (Well done, the UN, on the snappy title. Given the choice between a long winded but accurate title, or a geographically based title that gives no hint as to the actual content of the convention, the carriage industry always plumps for the latter, so of course the 2008 Convention is universally referred to as the Rotterdam Rules.)[14] The Rotterdam Rules made provision for the adoption of electronic bills of lading, but, of course, the Rules have never come into force, as they have yet to be ratified by the requisite 20 nations.[15]

Whatever the merits or otherwise of the Rotterdam Rules, which is of course a ripe topic for debate in and of itself,[16] the non-adoption of the Rotterdam Rules is perhaps indicative of the conservatism of the shipping industry generally and the resistance of anything that smacks of revolutionary change, particularly in relation to the implementation of a novel legal framework. In relation to this innate conservatism, it is pertinent to cite the words of DNV GL CEO Henrik O Madsen on his retirement in June 2015, when he said:

> Shipping is a very conservative, actually too conservative, industry. The attitude in the industry is mainly that any new regulation introduced is basically negative. I could hope that, going forward, they will change from seeing every new regulation as a risk to instead also thinking of a regulation as an opportunity.[17]

While that statement was made specifically in the context of regulatory change, it applies equally to developments or changes in relation to the adoption of electronic alternatives to the paper Bill of Lading. In 2014 BIMCO introduced the electronic bills of lading clause, which provides:

> BIMCO Electronic Bills of Lading Clause 2014
>
> (a) At the Charterers' option, bills of lading, waybills and delivery orders referred to in this Charter Party shall be issued, signed and transmitted in electronic form with the same effect as their paper equivalent.
> (b) For the purpose of Sub-clause (a) the Owners shall subscribe to and use Electronic (Paperless) Trading Systems as directed by the Charterers, provided such systems are approved by the International Group of P&I Clubs. Any fees incurred in subscribing to or for using such systems shall be for the Charterers' account.

---

12 The BOLERO electronic bill of lading system is operated by Bolero International, "Bolero" standing for Bill of Lading Electronic Registry Organization.

13 Accessible at www.bolero.net/wp-content/uploads/2020/04/Bolero-Insights-The-Bolero-Rulebook-NW.pdf (accessed 1 February 2021).

14 Because the signing ceremony took place in Rotterdam. Would Magna Carta have had quite the same impact if had been known as the Runnymede Compromise?

15 While precise figures are not easy to come by, it is to be doubted that the five countries that have ratified are responsible for much of the international carriage of goods by sea. They are Benin, Cameroon, Congo, Spain and Togo.

16 For a comprehensive critique of the Rules, see, T. Nikaki and B. Soyer, "A New International Regime for Carriage of Goods by Sea: Contemporary, Certain, Inclusive, and Efficient or Just Another One for the Shelves?" (2012) Berkeley Journal of International Law, 30: 303–348.

17 Cited at: https://shippingwatch.com/Services/article7817989.ece (accessed 1 February 2021).

(c) The Charterers agree to hold the Owners harmless in respect of any additional liability arising from the use of the systems referred to in Sub-clause (b), to the extent that such liability does not arise from Owners' negligence.

However, this clause, of itself, does nothing to drive forward the adoption of electronic bills: it merely provides a fairly vanilla framework for their utilisation if charterers so wish, with paragraph (c) putting all risks on charterers save for those arising from owners' negligence.

Perhaps of more significance has been the approval by the International Group of P&I Clubs of various electronic bill of lading solutions (now numbering six),[18] which include blockchain solutions, and, as from February 2020, in the CargoX blockchain document transfer platform, include a neutral or public blockchain using the Ethereum network. This, at last, may be seen as a significant stepping stone towards a wider acceptance of electronic bills.

2.6 Why the resistance to electronic bills of lading?

As set out in the previous section, it was over 30 years ago that the CMI first published a set of Rules for use with electronic bills of lading. Yet, despite the steps that have been taken, there has not been a widespread take-up in the use of electronic bills. It is striking that there do not appear to be any decided English law cases involved electronic bills. Indeed, a word search on the Westlaw platform for "electronic bills of lading" throws up a single hit, in a Chancery Division case,[19] in which Stephen Morris QC sitting as a deputy High Court Judge observed, in a case involving issues entirely unrelated to the carriage of goods, that:

> Whilst there has been debate in the context of electronic bills of lading and other electronic documents, the current state of the law has not developed to the point where something which exists in electronic form only is to be equated with a physical thing of which actual possession is possible.

Why has this been so? The innate conservativism of the shipping industry, which has been mentioned above, is undoubtedly a key factor. The resistance to change is perhaps best seen in concerns about fraud. Yes, of course, electronic systems can be the subject of fraud, in some cases not dissimilarly to the ways in which paper systems can. An example can be seen in the case of *Glencore International v MSC Mediterranean Shipping Co*[20] in which the bad guys were able to acquire electronic PIN codes to authorise collection of cargo, and the courts held that the carrier was in breach in delivering to the bad guys rather than the genuine Bill of Lading holders. There is a close analogy to a fraudster simply presenting a forged Bill of Lading to take possession. But, surely, it is perverse to object to a widespread implementation of a system of electronic bills of lading because of concerns about electronic hacking, when the paper-based system that is currently in place is itself so much more susceptible to fraud?

There are a number of other arguments, which in truth are all grounded in simple conservatism, inertia and "it's what we have always done"-ism, which do not withstand serious

18 ESS, BOLERO, e-title, GlobalShare, Wave and CargoX.
19 *Armstrong v Winnington Networks* [2012] EWHC 10 (Ch); [2013] Ch 156, at [51].
20 [2017] EWCA Civ 365; [2017] 2 Lloyd's Rep 186.

scrutiny. The more legitimate objection to the universal adoption of electronic bills of lading is much more fundamental. As analysed above, English law in relation to Bills of Lading, including in relation to the transfer of contractual obligations, is premised on the physical endorsement and transfer of a hard-copy original document.

Absent any new statutory regime, companies seeking to establish an electronic bill of lading platform are required to adopt an entirely different legal model to achieve by a different route that which the law achieves under COGSA 1992. All current so-called electronic bills of lading are not really Bills of Lading at all. They are the combination of a registry and a contractual construct. By way of example, the BOLERO Rule Book is based on a system of novation. The BOLERO eBL is really an entry on a central register operated by BOLERO. Parties who wish to use the BOLERO system all have to sign up to the Rule Book, which forms a legally binding agreement between all BOLERO users. This provides for the novation of the contract of carriage when a new holder of the eBL is designated on the register. The scheme of novation seeks to mirror certain aspects of the regime of COGSA 1992, so, for example, when the original shipper "transfers" the eBL his rights, but not his liabilities under the contract of carriage are extinguished. In contrast when a subsequent "holder" "transfers" the eBL both his rights and liabilities are extinguished.

There are a number of concerns which have been voiced about such systems. First, there may be worries about the security and integrity of a centrally based registry system. Second, the system is "closed" in the sense that the BOLERO eBL can only be used for transactions where all relevant parties have signed up to the Rule Book and have thus agreed to the novation scheme for the transference of contractual rights. This is of course in stark contrast to the existing paper-based system whereby anyone can become a lawful holder of the Bill of Lading and COGSA 1992 applies as a matter of law. Third, there is as yet simply no case law as to how the Rule Book will be interpreted and applied by the courts. The drafting style of the Rule Book is commendably straightforward and "plain English" in style. But that cannot disguise the fact that the wording will be tested and pulled apart by lawyers in myriad different factual circumstances if there is a wider take-up of the system and cases start to filter through to the courts. The absence of any case law will inevitably lead to greater litigation uncertainties than might be the case under the existing paper-based system where there is such a wealth of case law in this and other jurisdictions.

Put bluntly, current electronic bill of lading models based on central registers and pre-agreed novation terms constitute a fairly revolutionary change in terms of the legal framework governing the rights and obligations of all parties involved in the trading and carriage operations. And the shipping industry is not one that is swift to embrace revolutionary change.

2.7 The revolutionary technology of blockchain may allow evolutionary legal change

Blockchain technology, and perhaps in particular a document transfer platform built on a neutral, public blockchain network such as Ethereum, may provide a solution to the problems just identified. Professor Paul Todd in a thoughtful and insightful article has proposed possible models by which an open blockchain system might operate, without legal intervention; in particular either a model based on a chain of assignments and indemnities or a model based on novations.[21] He observes that while legislative change would be welcomed, it probably cannot be relied on.

However, surely the time has come for legislative intervention to cut through the industry inertia caused by the fear of a leap to a new legal framework. Primary legislation is probably not required. As noted above, COGSA 1992 already empowers the Secretary of State to pass regulations to make provision for the application of the Act to cases where an electronic communications network or any other information technology is used for effecting transactions corresponding to the functions of a paper Bill of Lading.

In a working paper in 2001, the Law Commission observed that:

> The absence of an electronic bill of lading, and the existence of adequate legal provision for contractual schemes, mean that there is no immediate need for domestic reform. There may be a need for reform in the longer term if an electronic bill of lading is created.[22]

Indeed, the Commission went further, saying:

> Technology may in the future be capable of providing the commercial world with a true electronic equivalent of a paper bill of lading. However, there is no working equivalent now. Nor, as we understand it, is there likely to be in the near future.

That was, however, 20 years ago, and the then "near future" is now in the past. It is submitted that public blockchain network technology can indeed provide a true equivalent of a paper bill of lading and that accordingly the need for reform, as envisioned by the Law Commission, is now upon us. It is clear that there is a widespread acceptance of the concept of ownership and possession of blockchain tokens. It would certainly come as a surprise to the "owners" of Bitcoins if they were to be told that they did not own an asset or were not able to transfer that asset to someone else. Perhaps more importantly than the subjective feelings of Bitcoin investors/speculators, there is increasingly legal recognition of blockchain tokens or cryptocurrencies as assets. For example, on 28 September 2018, in the case of *Elena Vorotyntseva v Money-4 Limited t/a Nebeus.Com, Sergey Romanovskiy, Konstantin Zaripov*[23] Birss J granted a freezing injunction over cryptocurrency assets noting:

> I should say no suggestion has been made by the respondents that the cryptocurrency that was given to them does not belong to the claimant. Nor is there any suggestion that cryptocurrency cannot be a form of property or that a party amenable to the court's jurisdiction cannot be enjoined from dealing in or disposing of it. I am satisfied that the court can make such an order, if it is otherwise appropriate.

A key analysis of the status of cryptoassets is set out in the UK Jurisdiction Taskforce (chaired by Sir Geoffrey Vos), *Legal statement on cryptoassets and smart contracts*, published in November 2019.[24] Having referred to the fact that traditional paper-based commerce uses documents of title to facilitate dealings in certain kinds of assets, and that the key feature of a document of title is that it confers ownership rights on the holder that

21 "Electronic Bills of Lading, Blockchains and Smart Contracts" (2019) International Journal of Law and IT Int J Law Info Tech, 27: 339.

22 www.lawcom.gov.uk/app/uploads/2015/09/electronic_commerce_advice.pdf (accessed 1 February 2021), at 24.

23 [2018] EWHC 2596 (Ch), at [13].

24 https://35z8e83m1ih83drye280o9d1-wpengine.netdna-ssl.com/wp-content/uploads/2019/11/6.6056_JO_Cryptocurrencies_Statement_FINAL_WEB_111119-1.pdf (accessed 1 February 2021).

pass with the physical transfer of the document, so that a transfer of the document effects a transfer of the asset, the Taskforce said[25] (emphasis added):

> Cryptoassets are not documents which can be physically transferred and so, as explained below, do not fit easily into the existing law on documents of title. *There is a functional analogy when they are used as tradeable tokens which represent off-chain assets, and it might be said that the token should be treated by the law in the same way that it would treat a document of title, so that a transfer of the token is effective to transfer the off-chain asset.* However, documents of title are only recognised as such under statute or where there is an established mercantile usage, neither of which (*at least presently*) applies in the case of cryptoassets.

This can certainly be interpreted as a statement that there should be no conceptual bar to regulations being made pursuant to section 1(5) of COGSA 1992 to make the Act applicable to blockchain-based tokens being used as electronic Bills of Lading. There is, as the Taskforce notes, a functional analogy between blockchain tradeable tokens and documents of title: all that is required is legislative recognition, which can be achieved by secondary legislation pursuant to section 1(5). There are a number of attractions to this approach. Most importantly, it does not require a fundamental reshaping of the legal framework, as, for example, the contractual construct of novation-based models do. Rather, it brings blockchain-based electronic bills into the existing legal fold.

There are additional benefits too. There is, again, widespread recognition of the inherent security and "unhackability" of public blockchains. The lack of reliance on a centralised register is another attraction. Further, blockchain-based bills of lading can be married to the introduction of so-called "smart contracts" – self-executing contracts which can significantly reduce transaction costs. Perhaps paradoxically, by this route, the revolutionary new technology of blockchain may facilitate a much more evolutionary move to the widespread adoption of electronic bills of lading because the existing legal frameworks can continue to apply. The time for section 1(5) of COGSA 1992 to be used to extend the application of the Act to blockchain-based electronic documents of title is upon us.

Postscript

The above chapter was written in 2020. It concluded by calling for the use of section 1(5) of COGSA 1992 to extend the application of the Act to blockchain-based electronic documents of title.

Since then, the Law Commission for England and Wales has published a consultation paper and draft Bill which, if implemented, aims to give certain electronic documents, including electronic bills of lading, the same effect in law as their paper counterparts.[26]

In relation to bills of lading, the proposed Bill would have much the same effect as that proposed in this chapter, but of course the Bill is much wider in scope, and requires primary legislation, as opposed the secondary legislation that could be used pursuant to the section 1(5) route.

The Law Commission's draft Bill lists seven documents which are the subject of the proposals: bills of exchange, promissory notes, bills of lading, ship's delivery orders, marine

25 Ibid, at para 112.

26 The Law Commission of England and Wales, *Digital assets: electronic trade documents – A consultation paper*, Law Com CP No 254. See further: https://www.lawcom.gov.uk/project/electronic-trade-documents/ (accessed 26 November 2021).

insurance policies, cargo insurance certificates, and warehouse receipts. The Law Commission names these "trade documents" and clause 5 of the draft Bill empowers the Secretary of State to add or remove documents within this list as trade practices continue to develop.

This chapter identified two key barriers to the widespread adoption of electronic bills of lading: the "innate conservatism of the shipping industry", and the implicit requirement in English law that a bill of lading must be in hard-copy form. It is the second of these that the Law Commission's proposals aim to solve. The Law Commission suggests that this implicit requirement is not a problem unique to bills of lading. Rather, it is rooted in a more fundamental issue with how English law understands the relationship between the concepts of tangibility and possession.

The problem, which the Law Commission names the "possession problem", lies in the presumption in English law that, for a thing to be possessable, it must be tangible. This presumption is now an express rule of English common law following the decisions of the House of Lords in *OBG v Allan*[27] and the Court of Appeal in *Your Response v Datateam*,[28] which held that intangibles are not capable of possession.

The Law Commission's solution to the possession problem is to identify what characteristics would make a trade document in electronic form amenable to possession. In total, they have identified three:

(1) It has an existence independent of both persons and the legal system: it is not a bare legal right.
(2) It is capable of exclusive control: the nature of the electronic document does not support concurrent assertions of occupation or use. This quality is sometimes described as "rivalrousness".
(3) It is fully divested on transfer: that is, if A transfers the document to B, A must no longer be able to access or use the document.[29]

If a trade document in electronic form meets these criteria, clause 1(3) of the draft Bill names it an "electronic trade document" and it is considered possessable. For instance, an electronic bill of lading that exhibits these three characteristics would be treated as a bill of lading as a matter of English law, and all existing common law and statutory rules would apply to it as they would to a paper bill of lading.

The scope of the Law Commission's proposals is far wider than the intended scope of section 1(5) of COGSA 1992 as it is not limited to documents used in maritime trade. The Law Commission has asked in its consultation paper whether the proposals should go further, for instance extending to documents such as bearer bonds and banker's drafts. In addition, their electronic trade documents project runs in conjunction with the Law Commission's "digital assets" project. The latter project asks whether the three characteristics identified in the electronic trade documents project should have a broader application, such that any digital asset which exhibits all three characteristics should be considered possessable in law.

27 [2007] UKHL 21.
28 [2014] EWCA Civ 281.
29 The Law Commission of England and Wales, *Digital assets: electronic trade documents – A consultation paper*, Law Com CP No 254, para 5.47.

However, the Law Commission's proposals are also, in one sense, narrower than the scope of section 1(5) of COGSA 1992. Section 1(5) applies to bills of lading, sea waybills, and ship's delivery orders. In contrast, the Law Commission's proposals omit sea waybills. The only exception that has been expressly recognised is a straight bill of lading (considered a form of sea waybill under the 1992 Act), which the Law Commission subsumes within clause 1(2)(c) of their draft Bill as a "bill of lading".

The Law Commission has addressed the omission of sea waybills in the consultation paper. According to the Commission, the draft Bill only aims to cover those documents which must be possessed in order to have legal effect, and sea waybills do not satisfy this requirement.

While the omission of sea waybills may seem like a significant divergence from section 1(5) of the 1992 Act, one might question whether including sea waybills in the draft Bill would have any demonstrable effect in practice. Sea waybills are almost exclusively used in electronic form, despite no formal legal recognition of an electronic sea waybill. The Law Commission certainly considers that the existing commercial use of electronic sea waybills largely obviates any need for their reform. As such, one might consider the Law Commission's draft Bill to have, in practice, the same effect as was intended by section 1(5) of COGSA 1992.

In short, this is a most welcome intervention, and the draft Bill, if passed, may finally facilitate a widespread move to the use of electronic bills of lading, bringing all the advantages set out above.

CHAPTER 3

Distributed ledger technology and commercial insurance

The beginning of a new era?

*Barış Soyer**

3.1 Introduction

Distributed ledger technology (DLT) is an emerging technology and there is a real buzz in several parts of the insurance industry, mainly generated by insurtech start-ups,[1] that it can dramatically alter the way insurance business operates. In simple terms, DLT is a database that is distributed across several independent computing devices (nodes), such that changes to the data are protected and managed by cryptography and consensus, ensuring that data cannot be improperly tampered with and that all parties have identical copies that can be considered as a reliable source of truth.[2] As such, many believe that in the insurance sector it can help to achieve efficiencies in risk pooling and claims management, and play a vital role in detecting and eliminating fraud.

The main purpose of this contribution is to evaluate whether DLT could have such a disruptive impact on underwriting commercial insurance contracts, in managing claims and even perhaps in utilising insurance as a financial asset for insurance companies themselves. As part of this analysis, it will be deliberated whether this technological development can be accommodated within the current legal rules and principles of private law, or whether it is essential to consider further regulation of insurance business to ensure that interests of all concerned (i.e., parties to the contract, service providers and the state) are protected. However, this analysis cannot be carried out in a vacuum. It is vital to consider to what extent the nascent technology has been developed and tested in an insurance setting by considering several ongoing projects,[3] and what direction computer scientists expect this technological revolution to take in future.

To this end, after a clarification of terminology and different types of relevant technology in Section 3.2, with the help of a hypothetical scenario that reflects the current thinking of those developing the technology in the insurance context in Section 3.3 we analyse how the new technology could be adopted in an ideal world. In Section 3.4, we consider the prospect of utilising this technology in commercial insurance. In this part, we consider

* **Director of Institute of International Shipping and Trade Law, Swansea University.**

1 Potential of losing their market dominance is a factor motivating large insurance companies also to invest heavily in testing the limits and viability of such technology. The main concern of large insurance companies is that lack of investment on their part in emerging technologies might give an advantage to small entrepreneurial tech companies in redefining insurance markets by utilising new products they have developed.

2 CRO Forum, *Insurance and Distributed Ledger Technology: A Risk Manager's Perspective*, (2019), at p. 5.

3 Even though insurtech start-ups have engaged in many projects over the last few years, it should be noted that the success rate is low, and only a handful of projects are ongoing and might be ready for adoption in the near future.

 DOI: 10.4324/9781003155195-3

not only the viability of using various aspects of this technology in insurance law, but also whether private law rules and mechanisms can provide adequate solutions to potential legal problems that might emerge, and whether the current regulatory framework is fit to support the development of this technology.

3.2 Various applications of DLT

There is a considerable degree of confusion as to the relationship between DLT and blockchain technology. In fact, blockchain is a variant of DLT which became famous as the technology designed to enable Bitcoin. Blockchain technology organises data in blocks linked to one another in a chronological manner and updates these entries using an append-only technology. As entries are added, the independent nodes (computers) validate these updates to ensure that they agree with the conclusion reached. This validation and agreement on a single copy of the chain is known as "consensus" and is typically conducted automatically by a consensus algorithm.[4] Once consensus has been reached, the chain updates itself and the latest, consented version is saved on each node separately.[5] The ledger can only be updated by such consensus; without it, altering or deleting previously entered data is impossible.

It is not essential that every DLT operates by adding blocks or chains to the ledger. Some recent DLT systems are used to store data in the form of graphs and tree structures. Therefore, it is not inaccurate to say that every blockchain is a distributed ledger, but not every distributed ledger is blockchain. Nevertheless, both systems are based on the principle of decentralisation and consensus among independent nodes. Unless otherwise stated, in this chapter the terms *blockchain* and *distributed ledger* will be used interchangeably.

Depending on who has a right of access, a distributed ledger (or a blockchain) can be classified as "public" or "private". A public distributed ledger is potentially open to anyone, whilst a private one is accessible only to pre-defined users. A public ledger could enable every user to be involved in the validation process of transactions on the ledger (permissionless public ledger) or only allow pre-defined users to validate transactions (permissioned public ledger). Similarly, only users who have special rights can validate transactions in a permissoned private ledger. The type of DLT to be employed depends on the needs of those utilising it. Each has its own advantages and disadvantages. For example, a public permissionless system might enable users to maximise the benefits of decentralisation and achieves a high level of transparency, but this will certainly mean slower speeds due to the fact that it will take longer to validate transactions.[6] Equally, a permissioned private DLT might restrict the number of users but will certainly work more efficiently and privately.

4 Each blockchain adopts a consensus mechanism that is used to determine how data is verified, how conflicting information is resolved and how agreement is reached on committing changes to the blockchain without the trusted centralised authority. For example, one way of doing this is by solving a cryptographic puzzle by brute computational force for a state change to be committed to the blockchain (a mechanism known as "proof of work"). Another consensus mechanism is by allowing trusted entities to vote on whether to commit the state changes to the blockchain (a mechanism known as "proof of authority").

5 CRO Forum, *Insurance and Distributed Ledger Technology: A Risk Manager's Perspective*, (2019), at p. 6.

6 This could be ideal when the DLT is used to make entries to an open registry (e.g., the Land Registry). This is currently contemplated in some countries.

Lastly, a few words are in order to illuminate the relationship between DLT and smart contracts. DLT can provide a platform for smart contracts to operate. A smart contract is an umbrella term for a self-executing contract in which the terms of agreement between the parties are completely digitalised and written in terms of lines of code. A code of this nature is often deployed and executed on a DLT network. In order to determine whether the conditions for the performance of a smart contract have been met, data (input) from outside the ledger will often be required. Such a data feed is known as an "oracle". Oracles provide the external data when needed, attest to the correctness of the data and push it onto the distributed ledger,[7] and accordingly play a key role in the processing of smart contracts.

3.3 How would DLT-based insurance work in an ideal world?

It has been envisaged that DLT, assisted by devices embedded with sensors and software that can connect to other devices and systems over the internet to exchange data,[8] could potentially provide a great opportunity for insurers to enhance their customer base, engage in more granular risk assessment and manage claims in a more efficient manner.[9]

Let us consider a potential product that makes use of such technologies and could potentially revolutionise home insurance. Imagine that a consumer looking for home insurance cover signs up to a platform supported by a blockchain web application. On sign-up, the customer is issued a public and private key, the latter of which is stored securely on his/her device or in his/her browser. The customer receives and is asked to install a tamper-proof device in his/her home. This device has a hardware-secured private key of its own, a GPS locator, a built-in camera, and the ability to detect and send information on various matters, such as the condition of fire and burglar alarms, water levels, condition of other devices plugged in at home, or temperature of the building, using its own private key. The data is encrypted and stored within a smart contract on the blockchain as part of the customer's account using his/her device's private key. This data is stored across each node of the network. For risk assessment purposes, the platform would also require the customer to provide relevant information such as the year the house was built, modifications to the property and his/her claim record. This personal data will be encrypted and stored on the blockchain.

The customer requests a quote for home insurance by permitting the insurers on the platform to access his/her device, collecting data on the condition of his/her house and other relevant matters (e.g., flood levels) and any personal information stored on the blockchain that may influence the quote. This quote would also be subject to the requirement that the customer keep the device originally provided in operation throughout the cover period. The customer then selects the best quote offered by insurers on the platform if (s)he wishes. The offer is confirmed by the customer signing the transaction with his/her blockchain private key. At this juncture a binding contract between the customer and the relevant insurer is established.

Several months later, the customer's house is flooded. The customer submits a claim through the platform. The insurer reverts to the data and images stored on the blockchain

7 CRO Forum, *Insurance and Distributed Ledger Technology- A Risk Manager's Perspective*, (2019), at p. 8.

8 A system of such devices is commonly known as the "Internet of Things" (IoT).

9 See A. Cohn, T. West and C. Parker, "Smart After All: Blockchain, Smart Contracts, Parametric Insurance and Smart Energy Grids" (2017) Georgetown Law Technology Review, 1: 273.

directly by the device, signed by the device's own hardware-secured private key. This time-stamped data, coupled with automated analysis of the flood level detector devices and public flood information confirming that a flood occurred in the area, provides irrefutable proof of an incident covered by the policy. The report is accessible only by the specific insurer who requires proof of the claim: no other customer or even the platform owner can access it. Once the claim is checked by the system against the details of the legally binding smart contract, the payment may be initiated automatically.

A DLT-fuelled revolution, or too good to be true? A careful eye must have already noted a serious limitation. The smart contract that facilitates payment could function in this example, as it is possible to obtain data from the device installed on the water levels in the property and check this data that can be objectively obtained from independent public bodies. But what if the loss was allegedly caused by burglary or theft? The device installed could have stored data on the condition of the burglar alarm, but what if there is a possibility that the culprits gained entry to the house because the customer left his/her door unlocked? Home insurance policies usually contain an exclusion for such losses. In that case, the information collected by the device might not be able to give a clear indication to the self-executing smart contract as to whether payment should be initiated or not.[10] Also, there could be an uncertainty as to the value of the goods stolen. Presumably in such cases, the insurers would wish to carry out a claim assessment anyway prior to initiating payment. This example clearly demonstrates that the value of smart contracts that can be used to manage claims will be very limited and can only provide benefit in instances where it is possible with the aid of independent data providers to establish the occurrence of the loss.[11]

The model just described could also create regulatory and legal difficulties. There is no doubt that the data uploaded on the blockchain by the customer will constitute personal data.[12] It is, therefore, essential that controls are put in place and the blockchain system is constructed in a manner so as to protect the personal data as required by law.[13] This might be technically possible by making alterations in the structure of the blockchain platform. The key point will be to ensure that once one of the insurers enters into a contract with the customer, the personal data of that customer is no longer available to other insurers. This is possible if the blockchain used is a permissioned one that is open to pre-defined users. A more difficult issue will be how the data subject's right to request correction or erasure of data, both of which are fundamental rights secured under the Data Protection Act 2018, can be accommodated given that such data is stored in an immutable blockchain. Computer scientists believe that there is a way around this if only the hash value of the personal data is stored on the blockchain. Given that it is probabilistically impossible to reverse-engineer a hash value and reveal the original anonymised data, it is arguable that this data does not constitute personal data that attracts the legislative controls imposed by the data protection

10 Unless, of course, the device provides 24-hour surveillance of the property; but that will undoubtedly be a massive intrusion of privacy, and it is hard to see many assureds agreeing to it.

11 In fact, there are examples of the technology being utilised successfully in the context of parametric insurance (e.g., flight delay insurance). This will be deliberated further at the end of this section.

12 Section 3(2) of the Data Protection Act 2018 defines "personal data" as "any information relating to an identified and identifiable living individual".

13 As far as the Uk is concerned, the relevant legal framework can be found in UK General Data Protection Regulation (UK GDPR), which is based on EU General Data Protection Regulation (Reg (EU) 2016/679), and the Data Protection Act 2018, designed to supplement UK GDPR.

legislation.[14] An alternative could be not to store any personal data on the blockchain but to use pointers on the blockchain to refer to where the personal data is stored off the blockchain, which can be readily deleted or amended. Insurers operating on the blockchain must also ensure that the personal information they hold with regard to the customer is not used for a different purpose (repurposing of data) without the consent of the customer.[15]

In the model described, the IoT device plays a significant role not only in terms of risk assessment but also in providing data to the assured and insurer with a view to informing them of any irregularity (e.g., malfunction of the fire alarm) and ultimately providing data that is required to execute the smart contract to pay a loss. The reality is that such devices could create vulnerabilities (to cyberattack, for example); or even increase the risk of loss (for example, by failing to report to the customer and the insurer a malfunction of the fire alarm as a result of a connection error). It is natural to expect the insurer to have a responsibility to ensure proper functioning and continuous monitoring of such devices. On that basis, there is no reason why the issue cannot be settled in the contract, by expressly allocating the risk of malfunction. In the absence of an express allocation of liability, it would be open to the assured to argue the existence of an implied term that makes the insurer liable; but certainly this is a matter that regulators might wish to keep an eye on.

Last but not least, one can see several problems associated with the use of the smart contract to manage claims. Before embarking on an analysis of the possible difficulties, a few words on the legal nature of smart contracts are in order. A smart contract is essentially the encoding or digital memorialisation of a contract or parts thereof. Put differently, they are simply technological tools for automating the performance of certain obligations under the contract (e.g., payment of an insurance claim in this context) rather than a source of obligations.[16] So in technical terms, smart contracts constitute unilateral transfer mechanisms or technologies automating certain types of contractual performance. On that basis, it is hard to see how principles of contract law would pose any obstacle to smart contracts – a point acknowledged by the UK Jurisdiction Taskforce in the Statement published in 2019.[17]

Be that as it may, the use of a smart contract to facilitate payment under a home insurance policy remains rather problematic. First, it is a genuine possibility that the code of the smart contract might diverge from the obligation contained in the policy.[18] This could arise due to a failure to correctly convert natural language into programming language, or simply as a result of a bug in the system. One would expect in such a case the natural language of the contract to prevail;[19] but unless the insurers accept that a mistake has taken place, the matter might need to be resolved by a court or arbitrator (or Financial Ombudsman Service). In

14 Institute and Faculty of Actuaries, *Understanding Blockchain for Insurance Use Cases* (2020), at 17.

15 Article 5(1)(b) and (c) of UK GDPR.

16 J. Cieplak and L. Leefatt, "Smart Contracts: A Smart Way to Automate Performance" (2017) Georgetown Law Technology, 1: 417, at 418.

17 UKJT, *The Legal Statement on the Status of Cryptoassets and Smart Contracts* (2019). A copy of the Statement can be found at www.judiciary.uk/wp-content/uploads/2019/11/LegalStatementLaunch.GV_.2-1.pdf (accessed 1 January 2021).

18 Errors in programming codes are to be expected. It has been stated in S. McConnell, *Code Complete* (2nd edn, Microsoft Press, 2004), at 521, that industry average experience is about 1–25 errors per 1,000 lines of code for delivered software.

19 Unless of course the contract contains a clause stipulating that the code comprises the entire agreement; but it is hard to see how this could be justified in the consumer insurance context. It remains a possibility that a consumer could claim that a term in the contract that gives priority to a computer code over the natural meaning of the words is a violation of the FCA Principles for Business (PRIN) Handbook PRIN. 2.1.1.1 of this Handbook

that case, the task of an arbitrator or judge is not an easy one. Technical expertise is required to assess whether the code reflects the natural language of the agreement or not. It might make sense to defer such disputes to competent extrajudicial dispute resolution, but for this to happen it would need to be agreed by both parties in the contract.

The second problem relates to the reliability of the data obtained from external sources (in the example above, the data obtained from public authorities on flooding levels) required to execute the obligation to make a payment under the insurance contract. Smart contracts are not inherently smart enough to determine the credibility of particular data feeds. This introduces a new challenge generally known as the "oracle problem" whereby the execution of smart contracts can be compromised by unreliable data. Assume, for instance, a situation where payment is executed as a result of inaccurate information supplied by oracles. Upon discovery of that, the insurer might take legal action to reverse the situation. It can certainly be argued that payment was the result of a mistake and the court might be required to intervene to reverse the situation; but there is, in any case, no doubt that a dispute of this nature will adversely affect the public image of such arrangements.

This brings us to the third, and probably the most acute, problem associated with the use of smart contracts in this context. It will be very difficult to expect any smart contract to be able to make an assessment in instances where a subjective analysis is required. For example, even though using the data recorded on the blockchain and confirmed by oracles suggests that the insured property suffered from flooding, it would be virtually impossible for the smart contract to determine how much flood damage was there on the second floor of an apartment building. More significantly, how realistic it is to expect legal concepts such as "reasonableness", "diligence" and "materiality" to be expressed in the conditional logic of a computer code? It is not fanciful to expect a computer code to encounter difficulties in implementing such concepts.

The analysis clearly illustrates that smart contract technology can only be employed in the insurance context in cases where it is possible for the parties to define the terms of the cover with precision so that those terms can be executed by a digital protocol. Put differently, although one sees difficulties in adapting this technology to car or home insurance, where losses can arise in a number of different ways and a claim analysis might be required in most instances to determine the validity of a claim, insurance products that can be restructured with the aid of firmer parameters might be suitable for a smart contract revolution. There is no doubt that parametric insurance policies offer such potential, for instance, and indeed there are already a few platforms that offer this kind of cover. Parametric insurance is a form of insurance where the payouts are determined not through a claims adjuster surveying the damage but based on objective measures, such as the departure times of planes or magnitude of weather-related events such as hurricanes. There are several parametric flight delay insurance products, for example, that operate on blockchain technology. Such products are usually offered online, and customers who wish to purchase such products provide personal data and flight data which are stored externally (probably to avoid privacy issues discussed above). Flight data and destination are processed internally by a proprietary algorithm that analyses historical data in order to evaluate the risk of delay and calculate the appropriate level of premium. Interaction with the blockchain begins after

reads: "A firm must conduct its business with integrity." In a similar vein, 2.1.1.6 requires that "a firm pay due regard to the interests of its customers and treat them fairly".

the customer chooses the coverage level and makes a payment. The policy is then appended to a blockchain though a smart contract. After the flight lands, the actual landing time is sent to the smart contract through the blockchain. This data is automatically retrieved using public flight data. In case of a delay, the customer receives a payment automatically on the credit card used for subscription.

3.4 Possible uses of DLT in commercial insurance – legal issues and solutions

The discussion in the previous part on the potential mass application of DLT brings us neatly to the focal point of this contribution. To what extent can DLT be utilised in commercial insurance – and if so, could it operate under the existing regulatory regime? Before reaching a verdict on this question, it is essential to make some preliminary observations on the matter, highlighting the limitations of blockchain technology in the light of the discussion in the previous part.

3.4.1 Limitations of DLT in the context of commercial insurance

Even though the legislation dealing with privacy-related issues will not cause a problem here, since the primary concern of relevant legislation is with personal rather than commercial information, it is submitted that there are significant practical, commercial and legal restrictions on the potential use of the DLT in commercial insurance practice. At the risk of sounding rather conventional, the author wishes to highlight the following restrictions of DLT in this context:

i) The nature of international commercial insurance business makes it rather difficult for a public blockchain platform to operate successfully, even with the aid of the IoT, for underwriting purposes. The risks proposed by assured are often multinational and rather varied, especially in the context of transport, energy and marine insurance; and for a reasonable risk assessment to be conducted by insurers, an ongoing dialogue between independent agents of the assured (i.e., brokers) and potential insurers is required. For example, to be able to offer loss of business cover to a ship operator or charterer, it is essential to find out details of its commercial operations, and the nature of its business and its clients. Naturally, such information cannot be obtained unless a dynamic risk assessment is undertaken. It is also worth keeping in mind that brokers play a vital role in explaining, particularly to potential assureds from a different jurisdiction, the scope and nature of disclosure expected of them by law. Put differently, unlike home and motor insurance, risks posed by commercial insurance are not homogenous, making it necessary for both parties to engage in an ongoing risk assessment exercise.[20]

ii) One might, of course, suggest that private blockchain platforms could be suitable to facilitate underwriting of more conventional risks, such as hull and cargo

20 It should be noted that the assured is under no obligation to make any disclosure in consumer insurance (see, Consumer Insurance (Disclosure and Representations) Act 2012) but disclosure is expected in procuring commercial insurance contracts.

insurance, with the aid of the IoT. In theory, this should be possible, but one can foresee several legal and commercial problems if a group of insurers were to offer such a product. The main commercial difficulty in this scenario will be the issue of who will own the blockchain platform. If it is owned by a few founding members, other insurers might have reservations about participating in the blockchain solution over concerns that they did not own the relevant IP rights. Of course, it is possible that a third party (independent from insurers) might own the platform and come up with a governance agreement that bound all participating members. The governance agreement, for example, could provide answers to issues like the legal responsibilities of participants if data was stolen as a result of a cyberattack.[21] The governance agreement could be subject to a particular law (e.g., English or French law) and also contain an exclusive jurisdiction clause. However, due to the blockchain ledger's non-specific location, transactions might potentially be subject to the jurisdiction of any given node. Private international lawyers will testify that even exclusive law and jurisdiction clauses can be subject to challenge. This potentially means that in certain jurisdictions courts might be willing to seize jurisdiction, given the lack of physical connection of the blockchain platform to any jurisdiction. And this prospect introduces a significant commercial uncertainty for insurers considering whether to utilise this new technology – unless, of course, there is trust and willingness among the participants to solve problems within the parameters of the corporate governance structure of the DLT platform.

iii) As discussed in the previous part, smart contracts will have a very limited utility in terms of claims management in instances where a thorough analysis of the causes and the extent of loss is required. This is often the case in transport, marine, aviation and energy insurance sectors. For example, if a cargo is damaged due to fire, which is usually an insured peril, there is still a need to assess the cause of the fire in case it is linked to an excluded peril. Also, in a case where the assured claims expenses incurred under a sue and labour clause, it is essential to determine whether the expenses incurred reasonably and whether they were incurred to minimise and/or avert an insured peril. It will be impossible to expect a smart contract to engage in such a subjective analysis.

3.4.2 Possible commercial applications of DLT in commercial insurance

3.4.2.1 Creating a secured network among counterparties involved in underwriting

It nevertheless remains a viable prospect to establish a private DLT platform for a limited number of insurers and their trusted brokers for the purposes of providing insurance cover for a specified type of risk (e.g., hull or cargo insurance). A well-informed broker could certainly add data required for risk assessment purposes to the blockchain (in the case of hull insurance, vessel's survey reports, claim history, information on the managers, etc.) for

21 Cracking of the encryption algorithms that underpin a DLT, although considered unlikely today, remains possible. In addition, technical evolutions (e.g., quantum computing) may render the security systems inherent to DLT systems irrelevant through time.

the information of insurers participating. It is vital that the presenting broker keeps an open dialogue with insurers interested in offering cover and continues bridging the information gap between the parties by responding to enquiries of insurers on the blockchain who are actively contemplating providing cover. The ongoing dialogue is essential in ascertaining the terms of the cover on offer. The ultimate purpose of this DLT is to enable insurers participating on the platform to subscribe to a proportion of the risk (e.g., to take 20–30% of the risk). It is imperative that once this is achieved, the access of insurers not involved in providing insurance cover for that risk is restricted for commercial reasons.

A commercial application of this nature has the potential to streamline underwriting processes by connecting brokers, insurers and assureds to a distributed common ledger that captures data about identities, risks and exposures and integrates this information with insurance contracts. This enables brokers (and their principals, ultimately) to reach quickly to a concentrated number of insurers who are in a position to consider providing instant cover for a particular line of business.

The position of the parties choosing to enter into an insurance contract by utilising the DLT in the manner described above is not altered in any fundamental manner under English law. The assured is still expected to present the risk fairly to the insurers under the Insurance Act 2015. A broker's failure to post accurate information on the platform with regard to an aspect of the risk, or provision of misleading answers with regard to the questions posed, will certainly have adverse legal consequences for its principal (assured).[22] For example, failing to provide accurate information might mean that the insurer could avoid the contract as long as this is deemed material,[23] and it is possible to show that, but for the breach, the insurer would not have entered into the contract of insurance at all.[24] The assured could potentially have an action against the broker under contract or in tort, especially if the broker took upon him/herself to present the facts or circumstances in a misleading manner.[25]

The advantage of this system is that it operates among trusted partners (insurers and brokers) so an effective governance structure can be put in place to deal with issues that might arise. Imagine a situation where due to a system error,[26] key information is not made available to the insurers for the purposes of risk assessment even though such information has been placed on the blockchain by the relevant broker. This presents interesting legal problems such as whether the insurer can be deemed to know such information. The relevant section of the IA 2015[27] provides no guidance on this matter, but assistance can certainly be drawn from case law decided prior to the introduction of the Act, as it was certainly not the intention of the draftsmen to make the principles emerging from those authorities

22 See ss. 3 and 7 of the Insurance Act (IA) 2015.

23 By virtue of s. 7(3) of the IA 2015 a circumstance or representation is material if "it would influence the judgment of a prudent insurer in determining whether to take the risk and, if so, on what terms".

24 Section 8 and Schedule 1 of the IA 2015.

25 See, for example, *Aneco Reinsurance Underwriting Ltd. v Johnson & Higgins* [1998] 1 Lloyd's Rep 565.

26 Although the DLT system is usually regarded robust, both the nodes and the corresponding clients running the DLT applications could be vulnerable to typical security risks, such as unauthorised access, virus attack, or missed security patches, with the consequence that those devices constitute vulnerabilities for the whole DLT network. For more info, see, CRO Forum, *Insurance and Distributed Ledger Technology: A Risk Manager's Perspective* (2019), at 16–17.

27 See s. 5.

redundant.[28] A fair number of legal authorities have indicated that information received by an agent of the insurer, who is authorised to take managerial decisions, is deemed to have been received by the insurer itself.[29] The legal issue here is whether the blockchain platform could be viewed in a similar manner so that information posted there for risk assessment purposes by the assured (or its agents) is deemed to have been received by the insurer even though that might not be the case in reality. It is submitted that by analogy one can plausibly argue that there is no fundamental difference between an agent of the insurer receiving information and failing to pass it on to the insurer, and information being appropriately placed on a blockchain but due to a system error the insurer is not able to access such information, especially given that it is the main promise of a DLT platform that such data will be accessible to all users. Assuming that this is the accurate legal analysis, an insurer will be precluded from taking any action against an assured for unfair presentation of the risk and potentially suffer financial losses. One would expect a governance regime to allocate the risk of, and liability for, such failures between the relevant parties and possibly the owner of the platform. The governance regime should also consider how the risks will be allocated between participants if it proves possible to crack the encryption algorithm as a result of the advances in quantum computing. The governance structure should also make allowance for maintenance requirements of the platform and costs associated with that. It might even be a sensible commercial stance to consider putting in place insurance arrangements to cover the liabilities and expenses incurred by the participants of the platform.

As the commercial model just described is designed to operate among trusted parties, one would expect those concerned to respect the governance and dispute resolution system put in place. Problems could arise if the assureds attempted to challenge the governance system in other jurisdictions; but in theory, as long as errors on the blockchain did not deprive them the cover provided under the insurance contracts concluded, it is hard to see on what grounds they could take action.[30] To conclude, it is submitted that such a commercial application can work efficiently and does not require the intervention of regulators. Any emerging problems can certainly be addressed within the parameters of private law. Put differently, the author does not see any regulatory vacuum here that should attract the interest of regulators.

3.4.2.2 Limited scope for smart contract applications

The commercial solution discussed here can be supported with a smart contract that is used to facilitate an extension to the cover. Imagine a hull insurance contract concluded between the assured and various hull insurers on the platform. In most traditional hull policies, the insured vessel is expected to operate within certain navigational limits. This cover can be extended to cover the vessel beyond the initially agreed navigational limits as long as notice is given to the insurer in a timely manner and an additional premium is agreed.[31] If the vessel

28 Recently, the Scottish appeal court, Inner House, Court of Session, in *Young v Royal and Sun Assurance Alliance plc* [2019] CCOH 32, stressed that the introduction of the IA 2015 did not alter the legal position with regard to waiver established by the case law pre-dating the Act.

29 *Tate v Hyslop* (1885) 15 QBD 368; *Bancroft v Heath* (1901) 6 Com Cas 137; *Equitable Life Assurance Sy v General Accident Assurance Corp* (1904) 12 SLT 348.

30 Perhaps this amplifies the need to have in place a liability insurance for the platform.

31 For example, cl. 2 of the Institute Time Hull Clauses (1/11/95) stipulates: "Held covered in case of breach of warranty as to . . . locality . . . provided notice to be given to the Underwriters immediately after the receipt of advices and any amended terms of cover and additional premium required by them be agreed."

is equipped with a GPS system that feeds data onto the blockchain platform as to the location of the insured vessel, a smart contract can execute an automatic extension to the cover and calculate additional premium payable on the vessel's return to the originally agreed limits. The GPS signals in this case provide reliable information required to execute an extension to the cover. Traditionally, when an extension to the existing cover is sought, the assured is under a duty of good faith relating to fair presentation of the extension request.[32] An interesting legal question is whether the assured here is expected to make further disclosure as part of that duty, or whether the fact that an automatic extension is granted when original geographical limits are exceeded is a kind of waiver on the part of the insurer of their expectation of any further disclosure from the assured. The author believes that the latter is the case here, since, taking the case law into account, it is evident that good faith duty is associated with the notice requirement, which is no longer needed when data coming from the GPS signal onto the blockchain performs the function traditionally performed by notice requirement. However, problems similar to the ones highlighted in the previous part could still arise. Imagine, for example, that the GPS signal is altered as a result of a cyberattack and sends inaccurate details as to the location of the vessel onto the blockchain, initiating the self-execution of the smart contract on the premise that the vessel has been outside the agreed geographical limits. The governance structure of the platform should provide a solution to such a problem, clearly stipulating that an executed smart contract could be reversed if the execution is the result of inaccurate data coming from the GPS.

This model could also be adjusted to work in the context of arranging reinsurance cover. A private and permissioned DLT can bring trusted insurers, reinsurers and brokers together, enabling the insurers and brokers to put details of risks (coverage terms for the original policy, experience data, etc.) they are seeking to cede on the DLT platform. This is made available to all potential reinsurers who could subscribe to these risks in the manner described above.[33] When it comes to claim handling for such reinsurance arrangements, a smart contract can be utilised to automate the execution of a reinsurance treaty. For example, on proportionate or quota sharing treaties, claims are reported in bulk on a periodic accounting basis.[34] A smart contract can be put in place to execute payment automatically when such data is shared on the DLT. Of course, the governance structure must enable a reinsurer to claim payments made back under a reinsurance treaty if reinsurers came to the conclusion that the reinsured's claim was not justified or excessive after performing an audit of the reinsured or examination of the claims files.

It is submitted that smart contract applications could potentially operate successfully in these contexts to reduce the admin costs. Especially if they are applied in managing claims for reinsurance treaties, this will reduce the need for manual records, address

32 *Overseas Commodities Ltd v Style* [1958] 1 Lloyd's Rep 546. The scope of the utmost good faith duty in this context has been clarified by Leggatt, LJ, in *Manifest Shipping & Co Ltd v Uni-Polaris Insurance Co Ltd & La Réeunion Européene (The Star Sea)* [1997] 1 Lloyd's Rep 360, at 370 (affirmed in the House of Lords [2001] UKHL 1; [2003] 1 AC 469, at [54], by Lord Hobhouse). See also s. 2(2) of the IA 2015, which stipulates that the Act applies in cases of any variation of the insurance contract.

33 The Blockchain Insurance Industry Initiative (B3i) deployed in 2017 such a DLT which provides opportunities to participants to enter into excess of loss reinsurance contracts for property catastrophe. This DLT has no concept of blocks or chain but rather handles data peer-to-peer transactions among participating partners.

34 Excess-of-loss reinsurance treaties operate in a similar way: claims generally are reported on a periodic accounting report or bordereaux providing claim-level detail about individual claims over a certain attachment point.

current inefficiencies and weaknesses in claim reporting, and simplify sharing of data like bordereau and claims databases. However, it should not be ignored that they can work only amongst a trusted group of partners who would act diligently and with integrity and respect the governance structure of the relevant DLT platform. It is hard to see how any smart contract solution could operate outside these parameters (i.e., a trusted group of insurers, reinsurers and brokers).

3.4.2.3 Securitisation of the DLT

Securitisation of insurance is a financial technique that consists in transferring insurance risks to investors operating the financial markets. This transfer is performed by gathering the risks and turning them into notes on the capital markets.[35] It has been suggested that DLT could help to make the securitisation process work in a more efficient manner by enabling the insurance-linked security to be packaged into a token (i.e., a digital representation of the insurance-linked security), potentially widening the investor base.[36] This is certainly a progressive idea; but apart from difficulties in constructing a reliable DLT platform that can accommodate insurers and potential investors, such a development will certainly attract the attention of financial sector regulators, given that it has potential for financial irregularity. In fact, this is a discussion that is a part of a larger debate on how the DLT technology should be regulated within the financial sector; but as far as securitisation of insurance is concerned, one feels that this is a rather premature prospect at this stage.

3.5 Concluding remarks

DLT technology can certainly provide some benefit to the insurance sector, and there is no doubt that insurers will consider utilising the new technology to the extent that there is a business case to do so. However, one should not assume that it will revolutionise risk pooling, undertaking and claim management in the commercial insurance (especially marine, aviation and transport) sector overnight. As discussed in this chapter, the DLT solutions could in the short run operate only between trusted participants to accelerate the underwriting process; and when it comes to reinsurance treaties, smart contract solutions could certainly bring efficiency to claim management systems. As long as DLT solutions operate among trusted partners, as illustrated in the paper, private law rules can provide solutions to legal problems that can arise. To be able to utilise DLT solutions more broadly, regulation will need to be put in place; but unless solutions accepted globally are found, which will be a very difficult thing to do, it is rather doubtful that participants would feel confident enough to engage in this new way of doing business.

35 Various mechanisms are used in the process of securitisation of insurance. For example, natural catastrophe swaps are used to swap floating payments attached to the occurrence of a claim against the fixed payment whose amounts and deadline have been predefined.

36 Institute and Faculty of Actuaries, *Understanding Blockchain for Insurance Use Cases* (2020), at 12–13.

CHAPTER 4

UNCITRAL model law on electronic transferable records

The missing link towards e-shipping?

*Olivier Cachard**

4.1 Introduction

From the discovery of movable type printing with Gutenberg, each generation has been claiming responsibility for discovering disruptive technologies. It is not only a matter of pride and self-achievement, but also a way to raise hard cash from investors, by creating enthusiasm for new markets. Obviously, the public blockchain promoters follow the path of their predecessors, even though blockchain is in reality a combination of techniques developed more than thirty years ago in connection with so-called public-key encryption systems. The disruption, if any, comes from the conjunction of an unprecedented calculation power with the input of a large amount of data. In an interesting parallel with the transhumanism movement, this has led to the creation of a new religious movement named "*dataism*", the "*data religion*"[1] that has gained many adepts. The promoters of "*dataism*" are now trying to spread it in a number of sectors. They have met considerable enthusiasm in the shipping community, which indeed has been considering the perspective of electronic bills of lading for more than thirty years.[2]

In this context, UNCITRAL has succeeded in keeping pace with this continuous evolution, through its legislative processes involving NGOs and businesses. Its latest achievement is the enactment of the Model Law on Electronic Transferable Records, adopted by the UN General Assembly on 7 December 2017 and supplementing an existing body of model legislation related to E-commerce dating from 1996 (the UNCITRAL Model Law on Electronic Commerce (especially Arts.16 and 17)) and 2001 (the UNCITRAL Model Law on Digital Signatures adapting legislation to public-key encryption systems). Later, the United Nations sponsored a Convention on the Use of Electronic Communications in International Contracts (the "Electronic Communications Convention") and the familiar Rotterdam Rules (especially Arts. 8–10).[3]

* *Institut François Gény, Université de Lorraine Honorary Dean, Avocat à la Cour, Arbitre*

1 Y. N. Harari, *Homo Deus, A Brief History of Tomorrow*, Penguin, 2015, at 429: "Dataism declares that universe consists of data flows, and the value of any phenomenon or entity is determined by its contribution to data processing".

2 CMI, *Rules for Electronic Bills of Lading* (2018); N. Ong, "Blockchain Bill of Lading and the Uncitral Model Law on Electronic Transferable Records" [2020] J.B.L. 202; N. Abdelatiff, "An Ethereum Bill of Lading under the Uncitral Model Law on Electronic Transferable Records" (2020) Maastricht Journal of European and Comparative Law, 27: 250; K. Maxen, "Electronic Bills of Lading in International Trade Transactions: Critical Remarks on Digitalisation and the Blockchain Technology" (2020) Coventry Law Journal, 25: 31.

3 M. Sturley, T. Fujita and G.vd Ziel, *The Rotterdam Rules: The UN Convention on Contracts for the International Carriage of Goods Wholly or Partly by Sea* (Sweet & Maxwell, London, 2010).

DOI: 10.4324/9781003155195-4

The Model Law on Electronic Transferable Records is intended to promote a legal framework ensuring the validity of the electronic transferable records that are traditionally considered to be necessary for international trade and international finance. In this respect, it is not sufficient to admit the validity of electronic contracts concluded between parties familiar with each other, either through long-standing business relations or through a closed electronic platform such as Bolero.[4] Here we have to go a step further and admit the validity of electronic contracts in an open environment, with the ultimate step being the acceptance of the validity of digital negotiable instruments. In this regard, the perspectives opened by the *Model Law* in the shipping sector depend on its scope.

The scope of the Model Law includes bills of exchange, promissory notes, consignment notes and warehouse receipts, as well as bills of lading. It is thus more extensive than that of the earlier Electronic Communications Convention, which specifically excludes "bills of exchange, promissory notes, consignment notes, bills of lading, warehouse receipts or any transferable document or instrument that entitles the bearer or beneficiary to claim the delivery of goods or the payment of a sum of money". By contrast, bills of lading certainly are within the scope of the new Model Law, at least if they are order bills or bearer bills. Whether straight bills are included is less clear. According to Para. 88 of the UNCITRAL *Implementation Guide* it seems they are not:

> The definition of "electronic transferable record" does not cover certain documents or instruments, which are generally transferable, but whose transferability may be limited due to other agreements, for example in the case of straight bills of lading. The definition of "electronic transferable record" should not be interpreted as preventing the issuance of those documents or instruments in an electronic transferable records management system (see also above, Para. 21). Substantive law should determine which documents or instruments are transferable.

This said, however, one can immediately understand how crucial this Model Law is for the shipping sector, as it may finally generalise the circulation of "genuinely transferable" electronic bills of lading, even between parties that are not acquainted. The Model Law promotes a uniform approach to the transferability of electronic records (see Section 4.2). Several ways to transpose this model legislation into national law will then be explored (see Section 4.3).

4.2 Uniform approach to transferability of electronic records

In the Gutenberg galaxy, both the validity of the *instrumentum* and its transferability were essentially dependent on trust in the original document as bearing the original information in writing, and the genuineness of the signatures of the initial parties and any transferees.[5] As a result, the issue of transferability depended on the form of the transfer and on the satisfaction of the conditions required for trust in what appeared on the face of the document. Today, however, the valid transfer of a record to the relevant transferee has itself become an issue, as a digital document can be easily replicated and

4 Bolero International Ltd, *Bolero Rulebook* (September 1999).

5 See W. Holdsworth, "Origins and Early History of Negotiable Instruments I" (1915) Law Quarterly Review, 31: 12, 13 ff.

sent to virtually anybody. The Model Law considers transferability to be a function, not a substantial concept (see Section 4.2.1). It therefore relies on a functional test (see Section 4.2.2), complying with the principle of (technical) non-discrimination (discussed in Section 4.2.3).

4.2.1 Transferability is a function, not a substantive concept

In the *M*odel Law, there is no conceptual definition of transferability, something that gives a civil lawyer an uncomfortable impression of circular reasoning. According to Art. 2, an "electronic transferable record" is an "electronic record that complies with the requirements of Article 10". This is later expanded by Art. 10 §1, providing that where the law requires a transferable document or instrument, that requirement is met by an electronic record if certain (specified) conditions are met. In reality, this is an attempt to avoid any overlap with the notion of "negotiability" which has been carved out by a number of national provisions. This is made clear in the UNCITRAL *Commentary* at Para. 20, which states: "The Model Law focuses on the transferability of the record and not on its negotiability on the understanding that negotiability relates to the underlying rights of the holder of the instrument, which fall under substantive law." This means that transferability under the UNCITRAL Model Law is a condition precedent to negotiability, a matter for the applicable national law.

The lack of any positive definition is explained by the fact that transferability of electronic records is grounded on the non-discrimination principle embedded in Art. 7, stating: "An electronic transferable record shall not be denied legal effect, validity or enforceability on the sole ground that it is in electronic form." As a result, the digitisation of documents cannot be regarded as an obstacle to the transferability of a record, transferability being one requirement to be examined with other substantive requirements established by the applicable law. The negotiability of a record thus requires transferability plus other substantive conditions. The approach to transferability is functional.

The legal and technical design of the suggested system is that from the issuance of the electronic transferable record, through successive transfers and up to the end of its legal validity, only one person should be entitled to request performance of the underlying obligations at any given moment. In a chain of successive indorsements, each indorsee has this exclusive right to demand performance from the time of the indorsement to him until he transfers it to someone else. This is meant to avoid "double spending".

4.2.2 Transferability relies on a functional test

The criteria of transferability are the following under Art. 10:

- (a) The electronic record contains the information that would be required to be contained in a transferable document or instrument; and
- (b) A reliable method is used:
    - (i) To identify that electronic record as the electronic transferable record;
    - (ii) To render that electronic record capable of being subject to control from its creation until it ceases to have any effect or validity; and
    - (iii) To retain the integrity of that electronic record.

These criteria are based upon two related underlying notions, *data integrity* and *singularity*, that are wider than the related concepts in national law.

The starting point of the UNCITRAL Model law approach is that an electronic transferable record shall be "unique". However, uniqueness has to be distinguished from the national concept of originality. According to Para. 82 of the *Explanatory Note* to the Model Law: "Uniqueness is a relative notion that poses technical challenges in an electronic environment, as providing an absolute guarantee of non-replicability may not be technically feasible." Uniqueness is thus a goal to be achieved: as the *Notes* continue: "The purpose of that notion is to prevent the circulation of multiple documents or instruments relating to the same performance and thus to avoid the existence of multiple claims for performance of the same obligation." As a result, uniqueness is the desirable result of the *control* exerted by the legitimate holder or by third parties. The notion of *control* should not be reduced to the national concept of *possession* which has a special signification in both civil and common law countries. To a certain extent, possession endows the possessor with substantive rights, while factual control only is a process leading to uniqueness. This makes it necessary to define control. Two aspects of the electronic record under scrutiny have to be controlled so as to reach uniqueness: *data integrity*, and *singularity*.

First, *data integrity* is to be understood in the meaning of the "UNCITRAL acquis", implemented into national legal systems through their enactment of the Digital Signature Model Law. According to Para. 100 of the *Explanatory Note* to the Model Law:

> The notion of integrity is an absolute one. It refers to a fact, and as such, is objective, i.e. either an electronic transferable record retains integrity or it does not. The reference to the reliable method used to retain integrity is relative since the assessment of the reliability of each method is to be carried out in light of the specific function pursued by the use of that method.

Data integrity refers to the fact that there was no possible alteration of the electronic record.

Second, the *Explanatory Note* goes on to say:

> The "singularity" approach requires reliable identification of the electronic transferable record that entitles its holder to request performance of the obligation indicated in it, so that multiple claims of the same obligation would be avoided. The "control" approach focuses on the use of a reliable method to identify the person in control of the electronic transferable record.

As a result, singularity does not refer to the electronic record itself (that may be easily duplicated in a digital format), but to the person in control of the transferable record, this person being the only one authorised to demand performance.

The concepts of *data integrity* and *holder singularity* both require indorsement to be achieved in an electronic way. According to Art. 15 of the Model Law, indorsement is to be made possible without any tangible support:

> Where the law requires or permits the endorsement in any form of a transferable document or instrument, that requirement is met with respect to an electronic transferable record if the information required for the endorsement is included in the electronic transferable record and that information is compliant with the requirements set forth in Articles 8 and 9.

This means that the rules relating to a valid endorsement must neither undermine the possibility of control of the data integrity, nor that of the singularity of the holder.

One question remaining is whether the execution of the indorsement has to be connected with the genuine civil identity of the endorsee, or whether it may be pseudonymous. The requirements for *data integrity* and *holder singularity* might plausibly be satisfied by the use of a pseudonym, on the assumption that the pseudonym is unique. But what if a claim has to be filed against an Indorsee? It then seems that it must be shown that the authenticated civil identity of the defendant corresponds with the pseudonym.

4.2.3 Transferability promoted through the principle of non-discrimination

The principle of non-discrimination is of paramount importance when it comes to rules about the transfer of electronic records. Obviously, the criteria of unacceptable discrimination are not those related to human rights law, but rather matters relating to either technical standards (Art. 7) or the country where the record originated (Art. 19). According to Art. 19 §1 of the Model Law: "An electronic transferable record shall not be denied legal effect, validity or enforceability on the sole ground that it was issued or used abroad." Para. 186 of the *Explanatory Note* fleshes this out by saying that the word "abroad" must be understood as being "used to refer to a jurisdiction other than the enacting one, including a different territorial unit in States comprising more than one". The prohibition on non-discrimination is crystal-clear: legal effect may not be refused to an electronic transferable record on the sole basis of its place of issuance or of use. But how does the non-discrimination principle operate in terms of positive obligations?

In the traditional public-key infrastructure contemplated by the 2001 UNCITRAL Model Law on Digital Signatures, the identity of the holder of a pair of encryption keys is to be certified by an approved national certification authority. This makes it possible to have mutual recognition of electronic signatures in different countries through the recognition of national certification authorities. However, in a distributed open technological environment, no such institutional mutual recognition seems possible. Two perspectives may be envisaged. One perspective would be to focus on the reliability of the technique used to issue and transfer the electronic record, regardless of the place of issuance or use. This perspective, how seductive it may be, leaves open the questions of what counts as a reliable system and who should assess its reliability. Another perspective would be to divide the world into two cosmogonies: nations that have enacted the Model Law (the cosmos) and those that have not (the chaos). Mutual recognition would be limited to transferable records issued and transferred in the forum, or in any other nation that had enacted the system. But this somewhat antique view of the "digital passport", however fascinating to apply, would also be very difficult to implement. This is because the Model Law is not an international convention: there are no states party, but merely a large variety of states who have drawn a total or partial inspiration from it.

The uncertainties surrounding the positive operation of the non-discrimination principle in an open environment may well explain that the drafters felt the need to include a conflict of laws rule in Art. 19 §2. This says: "Nothing in this Law affects the application to electronic transferable records of rules of private international law governing a transferable document or instrument." But this tribute to conflict of laws is somewhat vague. Is the transfer of the record a matter of formalism governed by the traditional rule *Locus regit actum*? Is the transfer a matter of substance? Substance may then equally relate to a contract or to a document of title. For the sake of legal predictability, it is suggested that a better solution would be to make one unique law applicable to the whole life cycle of a bill of lading qualifying as an ETR. But this immediately raises a new question: which law? Should it be the place where the bill of lading was issued, or the place where it was endorsed? Article 19 §2 is not an acceptable conflicts rule. It is more an indication of the limits of unification.

4.3 Ways for transposition to B/L in national laws

The UNCITRAL Model Law follows the principle of technological neutrality (see Section 4.3.1), which allows different options. However, the legal debate cannot be fully developed without due consideration for the forgotten question of the liability of the operators of the technical infrastructure (see Section 4.3.2). In terms of conceptual approach, one issue is to determine whether the autonomous functional concepts of the Model Law shall be merged with national law concepts for the transposition of the Model Law (III.C).

4.3.1 The principle of technological neutrality

The UNCITRAL Model Law, although adopted before the rise of blockchain, does not refer to any specific technology. Therefore, national legislators may opt between different scenarios consistent with the goals and functions defined by UNCITRAL. These scenarios may be divided into three main categories where the criteria of access, control, transferability and liability are not equally verified.

The first type of scenario would be a choice of a registry model known as a "centralized ledger", open to the contracting parties only. Access to the registry would be subject to the acceptance of the terms and conditions of the registrar, and would also require technical interoperability with the information systems of all parties to the contract of carriage. There is therefore neither a physical nor a digital possession of the electronic transferable record. However, the identification by the registrar of the person in control is functionally equivalent to possession. No direct transferability is possible between transferor and transferee; however, a transfer of control is performed through the intermediacy of the registrar. If the integrity of the electronic transferable record is altered, the liability of the registrar may be invoked, though subject to any limits contained in the terms of use. Essentially, this is the BOLERO scenario, with traditional cryptographic techniques.

The second type of scenario would be the choice of a "permissionless distributed ledger", open in this case to any party having internet access. Any operator in control of a pair of public and private keys could access the system, which thus would have no *ex ante* exclusion of fraudsters and rogues. There would be no physical possession of the electronic transferable record, but the holder would exert control over an electronic token. The strength of the algorithm would allow reliable identification of the electronic token, the uniqueness of which could thus be guaranteed. But this system suffers from a serious defect: it would offer no properly certified identification of the holder of the token. To a certain extent, this therefore resembles the situation of a bearer bill in a paper context. There would be direct transferability between transferor and transferee. Time stamping and cryptographic techniques could identify the earliest transfers and thus prevent "double spending". As far as any risk potentially affecting the transfer and the integrity of the token, there would be no legal action available to the victim. Essentially, the Permissionless Public Blockchain Scenario entirely relies on the technical capabilities of the system, with the belief in technical infallibility that is a dogma of "dataism".

The third type of scenario would be the choice of a "permissioned distributed ledger". This scenario strongly resembles the previous one as far as control, transferability and liability are concerned. However, access to the chain would be allowed only to operators able to meet a standard access criterion providing a relevant e-identity, for example through the national registries for commercial companies. Two varieties may be distinguished here: the Private Blockchain Scenario with limited access granted by a central authority, and the Consortium Blockchain Scenario based on consensus between a limited number of actors operating the nodes of the chain.

4.3.2 The forgotten question: the liability of the operators of the infrastructure

There seems to be a dogmatic trust in the robustness of the "distributed ledger" system, be it permissionless or permissioned. It is often asserted that no "double spending" is technically possible and that no loss of the token by the holder may be caused by the distributed ledger. As a result, it is assumed that with such a system the question of civil liability has become irrelevant. This supposed paradigm shift favouring risky operations without the remedy of civil liability is just one aspect of the global attempt to establish safe harbours for operators of IT systems: for operators of pipes and internet networks, for telecom operators as far as EMF emissions are concerned and for robots and artificial intelligence, applying to them a kind of legal personification and thereby treating them as if they were commercial companies. This model evolves from an individual civil liability model to the socialisation of risk through technical standards. Art. 12 of the Model Law thus considers technical parameters of reliability rather than possible civil liabilities.

This vision generates a juvenile enthusiasm among both marketers of public blockchain systems and also the general public, as if it were possible to squeeze out intermediaries through technical advances. The dark side of this reality, however, is that new intermediaries will replace traditional ones. First of all, the functioning of the blockchain requires "mining", that is, adding new blocks to the blockchain, an operation that requires the kind of heavy computing and human resources available largely to operators in China. This raises some strategic and sovereignty issues for maritime nations and shipping companies, as those controlling the "mining" process will also control the transfer of electronic records. Second, the identification process still requires the intervention of intermediaries. For sure, the blockchain makes it possible to guarantee that there is only one holder of a token at a time. But the digital identity of the holder has to be connected to a civil identity, which requires a certification authority. If the transfer of the record is done through digital endorsement, a pseudonym is enough to distinguish the transferor and the transferee. But what if the transferee needs to sue the transferor? Then the pseudonym is not enough and a certified civil identity has to be associated to the pseudonym.

4.3.3 The debate over national legislative techniques

There are different ways to enact the UNCITRAL Model Law. A state could simply require reference to the Model Law for interpretation when it comes to interpreting and implementing ordinary national legislation. Or specific legislation may be passed. Is it then necessary to alter or enlarge notions in national law so as properly to transpose the UNCITRAL Model Law? Questions arise, *inter alia*, about the notion of possession, with some authors submitting that the term "control" should be replaced by "possession".[6] Following this line of reasoning may lead to the enlargement or modification of fundamental concepts of national law. To a certain extent, in civil law countries, signature was not a legal concept before the introduction of digital signatures. In the paper world, signature was just a factual institution. Today it has gained conceptual density, as a legal definition was created to encompass both paper and digital signature.

6 E. Ong, "BlockChain Bills of Lading and the Uncitral Model Law on Electronic Transferable Records", NUS Law Working Paper 2018/020; Ch. Albrecht, "Blockchain Bills of lading: The End of the Story?" 43 Tulane Maritime L.J. 251 (2019).

CHAPTER 5

Autonomous systems

Cyber risks and seaworthiness

Paul Dean, Henry Clack** and Astrid Ainley****

5.1 Introduction

Seaworthiness is an ancient concept which is relevant for a number of aspects of maritime law, including charterparties, bills of lading and marine insurance. Breach by a shipowner of its obligations to provide a seaworthy vessel can have significant consequences, including potential liability to both charterers and cargo owners and the loss of insurance coverage.

Historically, shipowners have not considered cyber security to be a seaworthiness issue. Instead, it was seen as a business interruption risk which, whilst potentially costly, presented little threat to the physical safety of a shipowner's trading vessels. However, as vessels and shoreside operations continue to become increasingly reliant on both information technology systems ("IT") and operational technology systems ("OT") (i.e. real-world equipment which is managed by digitally controlled systems), there has been a significant increase in the industry's awareness of the risks posed by cyber issues, particularly following the 2017 NotPetya incident.[1] Nevertheless, the scepticism around the risk posed by cyber issues to the physical safety of vessels remains. Whilst the 2017 NotPetya incident is estimated to have caused $200–300 million of lost revenue for Maersk, there was no physical damage, loss of life or pollution. Similarly, both MSC[2] and CMA CGM[3] suffered varying degrees of business interruption as a result of malware attacks in 2020.

Maritime Autonomous Surface Ships ("MASS") are, of course, particularly prone to cyber issues. By their very nature, they are, first, heavily reliant on IT and OT, more so than a conventional vessel. Second, depending on the degree of automation displayed by the MASS at the time of the incident, it is likely to have less crew who, in the event of a failure of the technology on board, might be able to step in to prevent an accident. The IMO has defined the degrees of automation as follows:[4]

---

* **Partner at HFW; Global Head of Shipping; Head of HFW's Autonomous Vessel Group**

** **Associate at HFW; Member of HFW's Autonomous Vessel and Cyber Groups**

*** **Trainee at HFW, Member of HFW's Autonomous Vessel Group**

1 A. Greenberg, "The Untold Story of NotPetya, the Most Devastating Cyberattack in History" (*Wired*, 22 August 2018) www.wired.com/story/notpetya-cyberattack-ukraine-russia-code-crashed-the-world/ (accessed 1 May 2021).

2 G. Twining, "MSC Confirm Malware Attack" (*Safety at Sea*, 16 April 2020) https://safetyatsea.net/news/2020/msc-confirm-malware-attack/ (accessed 1 May 2021).

3 Safety at Sea, "Cyber Security" (2020) https://safetyatsea.net/category/news/news-cyber-security/ (accessed 1 May 2021).

4 International Maritime Organisation, "IMO Takes First Steps to Address Autonomous Ships" (25 May 2018) www.imo.org/en/MediaCentre/PressBriefings/Pages/08-MSC-99-MASS-scoping.aspx (accessed 1 May 2021).

 DOI: 10.4324/9781003155195-5

(a) "Degree One" means the MASS is operated with automated processes and decision support, with seafarers on board to operate and control shipboard systems and functions. Certain operations may be automated and/or be unsupervised at times but seafarers are on board ready to take control.
(b) "Degree Two" means the MASS is being remotely controlled with seafarers on board ready to take control and operate the shipboard systems and functions.
(c) "Degree Three" means the MASS is remotely controlled from a remote control centre without seafarers on board.
(d) "Degree Four" means the MASS is fully autonomous.

It is entirely possible that the degree of automation displayed by a MASS will change both over the lifetime of the vessel as new IT and/or OT is installed and, indeed, over the course of a single voyage. For example, it is possible that some port states might require that a MASS be crewed whilst it enters or leaves a port. However, on departure, once the MASS has reached outer port limits, the crew could disembark, leaving the MASS to carry out a long ocean voyage either under the control of a remote operator (Degree Three) or entirely autonomously (Degree Four). It should also be noted that, whilst advances are being made in respect of the automation of both MASS' navigational systems and other on-board operational systems, we are likely to see autonomous bridge systems before the entire maintenance regime can be carried out remotely for extended periods of time (i.e. without the presence of an engineering team on board). Seawater is, of course, corrosive and continuous maintenance is required to keep vessels operating. It is entirely possible therefore that we might see MASS navigating autonomously but carrying engineering teams.

Safety at Sea and BIMCO have run a maritime cyber security survey for the last five years, tracking the industry's awareness of and response to cyber issues. The 2019 survey demonstrated that there had been a notable shift in industry attitudes leading to a much greater understanding of the threats the industry faces.[5] This has led to an increase of 37% in the number of respondents who indicated that cyber security guidelines had been incorporated into their company or fleet. The Safety at Sea and BIMCO cyber security white paper's authors argue that this might explain the reduction in the number of respondents to the 2018 survey who reported that they had fallen victim to a cyber-attack within the previous 12 months compared to the 2017 survey (a reduction of 12% over the year).

However, it is also clear from the results of the 2019 survey that the respondents considered that the risk of OT being hacked was marginal, despite the fact that it could potentially cause an incident leading to damage to a vessel, a pollution event or even injury or death. This is surprising given the ongoing scrutiny of this issue by classification societies, research institutions and a range of different regulatory authorities. The Centre for Cyber Security ("CFCS") under the Danish Defence Intelligence Service has assessed that the collective threat from cyber-attacks against operational systems on Danish ships as being high, meaning that the CFCS estimate that, within the next two years, the operational systems on board one or several of the 700 Danish-flagged merchant ships will likely become the target of a cyber-attack.[6]

5 Safety at Sea and BIMCO, *Safety at Sea and BIMCO Cyber Security* (White Paper, 2019) 3.

6 Centre for Cyber Security, *Threat Assessment: The Cyber Threat Against Operational Systems on Ships* (1st edn, Danish Defence Intelligence Service, April 2020).

In 2019, only 42% of respondents answered yes to the question "Does your company protect your vessels from OT cyber threats?" Whilst there has been a significant improvement when compared to the 2018 survey results (when only 7% of respondents answered yes to the question), the 2019 results show that less than half of shipping companies have taken steps to protect their fleets from the risks posed by cyber threats to OT. Nevertheless, it appears that there is increasing industry awareness of the issue.

This increase in awareness is timely. The 2020 survey reported that 8% of respondents had said that cyber-attacks had resulted in system outages on vessels.[7] Furthermore, anecdotal reports in the 2020 survey revealed that respondents had experienced GPS spoofing and problems with ECDIS displaying incorrect information on vessels coming into berth. The US Coastguard has also reported that a large international ship experienced a successful malware attack that "significantly degraded the functionality of the on-board computer system".[8] Fortunately, essential control systems were not impacted.

Whilst there are likely to be a number of contributory causes to this attitude, there appears to be widespread scepticism of the threat posed to OT on board vessels with many in the industry believing that, due to the fact that many vessels are effectively "air gapped" from the outside world, the vessels are immune from cyber threats. This view is, in our opinion, potentially incorrect for a number of reasons, including the fact that improvements in technology mean that many vessel systems are now connected to external networks and the internet for maintenance or monitoring purposes. Furthermore, we continue to see incidents where either a vessel's crew or a third party (such as an agent, pilot or external service provider's technician) introduce malware into a vessel's systems.

In order to minimise the risk posed by cyber-attacks, many have recommended taking a multi-layered approach (founded on the principle of defence-in-depth).[9] This approach means that, even if one security measure is breached, the vessel, its crew (if any) and the wider environment are still safe. First, vessels should be protected by restricting physical access to their IT systems. This can include access cards, locked cabinets or maintenance scans. Crew training is key in order to prevent these protections be circumvented, intentionally or otherwise. Second, network security should be implemented using firewalls and data diodes in such a way so as to segregate different zones of the ship. This means that, even if someone were to gain physical access to a vessel's IT system, access to safety critical systems, such as the bridge, is prevented. Finally, individual system security controls should be in place so that, in the event of an attack, access to systems is limited and recovery options are available. Examples of this include, encryption, user control and authentication, together with back-up and recovery measures. Ideally, each of these layers should be borne in mind during the design phase of the vessel, before the keel is even laid.

As touched on above, seaworthiness is relevant in the context of marine insurance, charterparties and bills of lading.

7 Safety at Sea and BIMCO, *Safety at Sea and BIMCO Cyber Security* (White Paper, 2020).

8 United States Coast Guard, *Marine Safety Alert, Inspections and Compliance Directorate* (USCG 06–19, 2019).

9 Safety at Sea and BIMCO, *Safety at Sea and BIMCO Cyber Security* (White Paper, 2020).

5.2 Contracts of carriage

All contracts of carriage contain an implied obligation to provide a seaworthy vessel. Mr Justice Field summarised the obligation as follows in *Kopitoff v Wilson*:[10]

> the shipowner is, by the nature of the contract, impliedly and necessarily held to warrant that the ship is good, and is in a condition to perform the voyage then about to be undertaken, or, in ordinary language, is seaworthy, that is, fit to meet and undergo the perils of the sea and other incidental risks to which she must of necessity be exposed in the course of the voyage.

The classic definition of seaworthiness is that of Scrutton L.J. in *F.C. Bradley & Sons Ltd v Federal Steam Navigation*:[11]

> The ship must have that degree of fitness which an ordinary careful owner would require his vessel to have at the commencement of her voyage having regard to all the probable circumstances of it. Would a prudent owner have required it to be made good before sending his ship to sea, had he known of it?

If the answer to the question posed in the last sentence of that quotation is "yes", then the ship is unseaworthy.[12] The obligation to provide a seaworthy vessel is an absolute one and, at common law, a shipowner will be liable irrespective of fault if a vessel is unseaworthy. When determining whether or not a vessel is seaworthy, one commonly used test is whether a prudent owner would have required that the defect should have been made good before sending his vessel to sea, had he known of it. The test is therefore an objective one. It should also be noted that the standard of seaworthiness required is a variable one, depending on the nature of the voyage to be undertaken, the type of vessel and cargo carried and the likely dangers to be encountered. For example, a handy-size bulker may be perfectly seaworthy undertaking voyages in Russian waters in the summer but may be unseaworthy if it were to undertake the same voyage in the dead of winter when sea ice was present (unless it was an ice class vessel).

The obligation to provide a seaworthy vessel includes the obligation to ensure that the vessel is suitably manned (i.e. with a competent crew)[13] and equipped. Vessels have been found to be unseaworthy for a number of reasons, including defective engines or navigation equipment,[14] incompetent crew and where the documentation carried on board is

---

10 (1876) 1 QBD 377 (QB) 380.

11 (1926) 24 Lloyd's Rep 446 (CA) 454.

12 *McFadden v Blue Star Line* [1905] 1 KB 697 (KB) 706; and *Alize 1954 v Allianz Elementar Versicherungs AG (The CMA CGM Libra)* [2019] EWHC 481 (Admlty), 1 Lloyd's Rep 595 [75].

13 The incompetence of the master or crew may consist of (among other things) "disabling want of skill or disabling want of knowledge" or a "disabling lack of will on the part of a member of the crew to use such skill and knowledge as he possesses": *Owners of Cargo Lately Laden on Board the Makedonia v Owners of the Makedonia* (The Makedonia) [1962] 1 Lloyd's Rep 316 (QB) 334–335; or, as it was put in *Papera Traders Co Ltd and others v Hyundai Merchant Marine Co Ltd and another (The Eurasian Dream)* [2002] EWHC 118 (Comm Ct), [2002] 1 Lloyd's Rep 719, 737, "a disinclination to perform the job properly" The earlier authorities referred to above prefer to use the term "inefficiency of the crew" however the writer does not consider that there is a meaningful distinction between "inefficiency" and "incompetence".

14 The absence of appropriate and / or up to date charts can render a vessel unseaworthy. *The Marion* [1984] AC 563 (HL); *Alize 1954 v Allianz Elementar Versicherungs AG (The CMA CGM Libra)* [2019] EWHC 481 (Admlty), 1 Lloyd's Rep 595 [79]; and on appeal at [59]: "*it is well-established that an aspect of the vessel's seaworthiness is that she has the appropriate, up-to-date charts on board*".

inadequate.[15] There must also be adequate systems or procedures to deal with ordinary incidents which occur during a voyage.[16]

Whilst many of these historic examples pre-date the modernisation of the world's merchant fleet and the shipping industry's increasing reliance on IT and OT, comparisons can easily be drawn between these historic instances of unseaworthiness and the issues around cyber security that the shipping industry is currently facing. Engine or scrubber management systems may fail due to malware. GPS spoofing (whereby GPS information is falsified) may result in navigational errors and, ultimately, collisions and groundings. A lack of adequate cyber security training for a vessel's crew may result in the introduction of malware and mismanagement of a crisis situation may exacerbate losses. Finally, if certain port states were to impose a cyber security certification requirement for all vessels calling at their ports and a vessel were to do so without carrying the proper paperwork, then the vessel is likely to be found to be unseaworthy.

Seaworthiness also includes an element of cargoworthiness (i.e. the vessel must be in a fit state to receive the specified cargo). It is possible to imagine a situation where some form of malware disrupts a container vessel's power management system, leading to a failure of power supply to the reefers on board. If the reefers on board contained valuable, temperature-sensitive pharmaceutical products then substantial claims could be anticipated. Almost all charterparties revise the implied obligation to provide a seaworthy vessel by including express terms which introduce an obligation to exercise due diligence to ensure that the vessel in question is seaworthy. For example, the GENCON 1994 form provides that:

> The Owners are to be responsible for loss of or damage to the goods or for delay in delivery of the goods only in case the loss, damage, delay has been caused by personal want of due diligence on the part of the Owners or their Manager to make the Vessel in all respects seaworthy and to secure that she is properly manned, equipped and supplied, or by the personal act or default of the Owners or their Manager.

Similarly, a carrier under a bill of lading at common law has an absolute obligation to provide a seaworthy vessel, subject to a small number of exceptions (such as acts of God or inherent vice). The Hague-Visby Rules, which are incorporated into most bills of lading, replace this with an obligation to exercise due diligence.

Article III provides that:

1 The carrier shall be bound before and at the beginning of the voyage to exercise due diligence to:
   (a) make the ship seaworthy;
   (b) properly man, equip and supply the ship;
   (c) make the holds, refrigerating and cool chambers, and all other parts of the ship in which goods are carried, fit and safe for their reception, carriage and preservation.

15 Defects in a vessel's passage plan will render a vessel unseaworthy: *Alize 1954 v Allianz Elementar Versicherungs AG (The CMA CGM Libra)* [2019] EWHC 481 (Admlty), 1 Lloyd's Rep 595 [78]; and on appeal [72], [79] and [87].

16 *Compania Sud Americana De Vapores SA v Sinochem Tianjin Import and Export Corporation (The Aconcagua)* [2009] EWHC 1880 (Comm Ct), [2010] 1 Lloyd's Rep 1 [367]; *Alize 1954 v Allianz Elementar Versicherungs AG (The CMA CGM Libra)* [2019] EWHC 481 (Admlty), 1 Lloyd's Rep 595 [86]; and on appeal at [71].

2 Subject to the provisions of Article IV, the carrier shall properly and carefully load, handle, stow, carry, keep, care for, and discharge the goods carried.

Under this rule, a carrier has two fundamental obligations. First, under Article III, Rule 1, a carrier is bound before and at the beginning of the voyage to exercise due diligence to make the ship both seaworthy and cargoworthy. Second, under Article III, Rule 2, a carrier is required to properly load, stow carry, keep, care for and discharge the goods. Notably, the obligation under Rule 2 is a continuing one (i.e. it lasts throughout the whole time that the cargo is on board), unlike the Rule 1 where due diligence only needs to be exercised at the beginning of a voyage.

If a carrier is able to prove that he has satisfied its obligations under both Article III, Rule 1 and 2 then they are entitled to rely on Article IV, Rule 2 which includes a number of exceptions, such as fire and perils of the sea, however these are less relevant in the context of cyber security and we have therefore not referred to them. The exceptions that are likely are be relevant are:

(a) Act, neglect, or default of the master, mariner, pilot, or the servants of the carrier in the navigation or in the management of the ship.

. . .

(p) Latent defects not discoverable by due diligence.

. . .

(q) Any other cause arising without the actual fault or privity of the carrier, or without the fault or neglect of the agents or servants of the carrier, but the burden of proof shall be on the person claiming the benefit of this exception to show that neither the actual fault or privity of the carrier nor the fault or neglect of the agents or servants of the carrier contributed to the loss or damage.

Under these exceptions, if a carrier has exercised due diligence, then it will absolved of liability. Further, where loss has resulted from unseaworthiness at the beginning of a voyage and where has been a want of due diligence on the part of the carrier, the carrier is nonetheless entitled to rely on (p) if the want of due diligence was not causative of the loss.[17] It should, however be noted that (p) cannot be seen to be a true exception because the logic is somewhat circular (i.e. if the owner has exercised due diligence then it will not be liable for losses arising out of a defect which could not have been discovered).

Due diligence is equivalent to the exercise of reasonable care and skill; the standard of due diligence is similar to the common law duty of care such that, the lack of due diligence is equated with negligence. Likewise, due diligence does not require the taking of steps which are reasonably seen to have been necessary only with the benefit of hindsight. It is well-settled that the duty under Article III rule 1 to exercise due diligence is non-delegable – if any servant or agent fails to exercise proper skill and care in making the vessel seaworthy, the carrier is liable for the loss and damage resulting from that unseaworthiness.

17 *Kuo International Oil Ltd and others v Daisy Shipping Co Ltd and another (The Yamatogawa)* [1990] 2 Lloyd's Rep 39 (QB).

When it comes of the burden of proof, a claimant must prove that any unseaworthiness that is established was a cause of the casualty. However, it need not be the sole or even the dominant cause.[18]

Papera Traders Co Ltd and others v Hyundai Merchant Marine Co Ltd and another (The Eurasian Dream)[19] was a case where the owners of a vessel were held to have failed to act as a reasonably prudent owner and found to have breached their due diligence obligations. The case involved a fire on board a car carrier in Sharjah in July 1998 which was likely to have been caused by the charterers' stevedores who were refuelling and jump-starting vehicles at the same time as using batteries on a pick-up truck. At the time of the incident, the ship was discharging cars and bunkering. When the fire broke out, the crew and stevedores on board unsuccessfully started to try to fight the fire with extinguishers and hoses. When it became clear that these methods were not working, the vessel's CO_2 system was released but the crew had failed to make sure that the ventilators, dampers and gas tight doors were closed. The fire subsequently spread, destroying the cargo on board and causing the vessel to become a constructive total loss.

Ultimately, the English High Court held that the vessel had been unseaworthy for numerous reasons, including:

- The officer supervising the stevedores should not have allowed the cars to be refuelled and jump started at the same time.
- There were insufficient walkie-talkies and the fire extinguishers on board had not been properly serviced.
- The CO_2 system had not been properly deployed.
- The manuals supplied by the vessel's technical managers failed to give guidance as to the supervision of stevedores.
- The crew were improperly trained and there had not been any effective emergency drills carried out.

The causes of unseaworthiness in the case of the *Eurasian Dream* could, quiet easily, equally apply to a situation on board a modern vessel. For example, poorly supervised stevedores, port officials or charterers' agents might introduce malware into a vessel's computer network via an infected USB stick. Inadequate or out-of-date anti-virus software might prevent an infection from being promptly identified, allowing it to spread. Finally, improper guidance, documentation and training from or by the vessel's technical managers may result in the impact of a cyber-event being exacerbated potentially, if the vessel's OT is involved, leading to a pollution event, physical damage to the vessel or loss of life.

Whether or not a vessel is considered to be seaworthy will depend, to a large extent, on the prevailing practice of the shipping industry at the time of the voyage. Further, ships do not need to be fitted with the latest, cutting-edge technology as long as the general practice at the time of the voyage was not to adopt it. For example, in the 1960s, whilst radar or loran were available, it was not necessary for a vessel to be fitted with them in order to be considered seaworthy. This is obviously not the case nowadays.

18 *Northern Shipping Co v Deutsche Seereederei GmbH and others (The Kapitan Sakharov)* [2000] 2 Lloyd's Rep 255 (CA) 269–270.

19 [2002] EWHC 118 (Comm Ct), [2002] 1 Lloyd's Rep 719.

There has been an increasing amount of industry guidance regarding cyber security in recent years. Since 1 January 2018, the OCIMF's SIRE Programme requires tanker owners and operators to address cyber security risk in their policies and procedures, including having an internal cyber audit program, retaining independent cyber specialist support and updating vessels' ISM System/SMS and ISPS ship security plans to address cyber security risks.[20] Further, Rightship's Inspection and Assessment Reports[21] now contain a cyber security section containing a Safety, Security and Environmental management checklist. Finally, BIMCO continues to publish and update its Guidelines on Cyber Security Onboard Ships, the third version of which was published in December 2018.[22] In addition to BIMCO, the Guidelines are supported by CLIA, International Chamber of Shipping, Intercargo, InterManager, Intertanko, IUMI, OCIMF World Shipping Counsel and provide guidance in relation to identifying vulnerabilities, assessing risk exposure, developing protection and detection measures, establishing contingency plans and responding to cyber security incidents. In light of all of this guidance, it is fair to say that the shipping industry as a whole should be aware of cyber issues by now.

Despite this increased awareness, it is not clear, however, whether an aggrieved charterer or cargo owner under a bill of lading will be able to successfully bring a claim for damages against a shipowner if they have suffered a loss following a cyber incident. This is because, whilst industry awareness is increasing year on year, the guidance itself is not intended to proscribe precisely what steps all shipowners should be taking to protect against a cyber-attack. In fact, quite the opposite is true and Version 3 of the BIMCO Guidelines expressly state:

> The guidelines are not intended to provide a basis for, and should not be interpreted as, calling for external auditing or vetting the individual company's and ship's approach to cyber risk management.[23]

The situation at hand is similar to that which arose following the publication of the Best Management Practices to Deter Piracy in the Gulf of Aden and Off the Coast of Somalia ("BMP") in the mid- to late 2000s. Just as has happened with cyber security, the shipping industry reacted to a new threat (piracy) by publishing guidance in the hope of minimising losses. The publication of the BMP gave rise to a number of claims in which cargo owners argued that they were not liable to contribute in general average for the (often substantial) costs of freeing both captured vessels, cargo and their crew on the basis that the vessel had been unseaworthy due to the owners' alleged failure to properly implement the BMP. These claims, for the most part, failed because compliance with the guidance set out in the BMP was not mandatory. In fact the third version of the BMP was prefaced with:[24]

> IMPORTANT: THE EXTENT TO WHICH THE GUIDANCE GIVEN IN THIS BOOKLET IS FOLLOWED IS ALWAYS TO BE AT THE DISCRETION OF THE SHIP OPERATOR AND MASTER.

20 Jim Textor, 'New OCIMF pre-fixture tanker vetting cyber requirement' (*Gard*, 22 November 2017) www.gard.no/web/updates/content/24478791/new-ocimf-pre-fixture-tanker-vetting-cyber-requirement (accessed 30 April 2021).

21 RightShip, *RightShip Inspection and Assessment Questionnaire* (RightShip, April 2019).

22 BIMCO, *The Guidelines on Cyber Security Onboard Ships, Version 3* (*BIMCO*, December 2018).

23 BIMCO, *The Guidelines on Cyber Security Onboard Ships, Version 3* (*BIMCO*, December 2018) 1.

24 OCIMF, *BMP3 Best Management Practice 3: Best Management Practices to Deter Piracy Off the Coast of Somalia and in the Arabian Sea Area: Piracy Off the Coast of Somalia and Arabian Sea Area* (Witherby, 2010).

On balance, the present level of cyber knowledge in the shipping industry is such that it would arguably be difficult for an owner to successfully argue that a vessel's seaworthiness did not include an element of cyber security. However, if the shipowner did take certain steps (such as conducting a risk assessment and providing crew training), then cargo owners and charterers may face an uphill battle to successfully recover their losses.

By contrast, the Polar Code,[25] which entered into force on 1 January 2017 after the IMO adopted the safety measures in Part I of the Code (the pollution prevention measures in Part II and amendments to SOLAS, STCW and MARPOL), took a much stricter approach. The Code imposed the requirement for ships intending to operate in the Arctic and Antarctic to apply for a Polar Ship Certificate which was contingent on an assessment of the anticipated operations and hazards the ship may encounter in polar waters. The Code also introduced certain training requirements for seafarers, together with new rules around ship design, stability, fire safety, voyage planning and communication. The mandatory nature of these requirements means that charterers or cargo owners will find it easier to successfully argue that a vessel was unseaworthy for failing to comply with the Polar Code.

This situation is likely to change after 1 January 2021 because of IMO resolution MSC.428(98).[26] The resolution, adopted in June 2017:

> AFFIRMS that an approved safety management system should take into account cyber risk management in accordance with the objectives and functional requirements of the ISM Code.
>
> ENCOURAGES Administrations to ensure that cyber risks are appropriately addressed in safety management systems no later than the first annual verification of the company's Document of Compliance after 1 January 2021

This language is, however, not prescriptive at all and it will be up to each of the flag states to determine what is meant by ensuing that cyber risks are "appropriately addressed" in the safety management system. We therefore foresee that there is likely to be a range of responses, varying from an annual risk assessment and training for crew through to a requirement for continuous monitoring and extensive preparations. However, only time will tell and, as cyber incidents continue to affect the shipping industry (such as the recent attack on CMA CGM),[27] the industry will be further incentivised to improve standards.

5.3 Marine insurance

There is a good deal of uncertainty around how liabilities will be treated in the context of MASS. Some have argued that the traditional, fault-based, liability regime should continue to apply. Others have suggested that a strict liability regime, similar to the one being discussed for driverless road vehicles, should apply. For example, a report published by the Danish Maritime Authority concluded that:

> In connection with fully autonomous ships [Degree 4 – i.e. fully autonomous], there is reason to presume that it will not make sense to refer to liability based on fault to the extent that the

25 International Code for Ships Operating in Polar Waters [Polar Code], 2017, IMO MEPC 68/21/Add 1, *entered into force* 1 January 2017.

26 Resolution MSC 428(98) – Maritime Cyber Risk Management in Safety Management Systems, 2017, IMO MSC 428(98), *entered into force* 16 June 2017.

27 Safety at Sea, "Cyber Security" (*Safety at Sea*, 2020) https://safetyatsea.net/category/news/news-cyber-security/ (accessed 30 December 2020).

> navigation is performed and decisions of importance to the ship's course and speed are taken by an autonomous system without any human interference. It must be presumed that this could, in the longer term, change the liability norm, at least in connection with collisions, to strict liability on behalf of the shipowner.[28]

However, until a strict liability regime is agreed at national and/or international levels, owners of autonomous vessels will still have fault-based liability.

Furthermore, a report published by the Institute of Marine Engineering, Science & Technology (IMarEST), in conjunction with Clyde & Co, showed that approximately two-thirds of global marine industry executives believe there is uncertainty surrounding liability issues relating to MASS.[29] The situation is not helped by the fact that there is no universal definition of cyber risks. For example, the UK P&I Club has defined cyber risk as "the risk of loss or damage or disruption from failure of electronic systems and technological networks".[30] Conversely, the Japan P&I Club has described cyber risk as "is a potential risk . . . which will cause financial loss, disruption or damage to the reputation of an organization" (i.e. not including the risk of physical damage to a vessel).[31]

Traditionally, losses arising from a cyber-attack have been excluded from cover due to the inclusion of the Institute Cyber Attack Exclusion Clause 380 which provides:

> in no case shall this insurance cover loss damage liability or expense directly or indirectly caused by or contributed to by or arising from the use or operation, as a means for inflicting harm of any computer, computer system, computer software program, malicious code, computer virus or process or any other electronic system.[32]

In order for an insurer to avoid coverage, it must prove that the loss occurred as a result of a malicious act. The most recent authority on this point is the decision from the Supreme Court in *The B ATLANTIC*.[33] In this particular case the Supreme Court was asked to determine whether the attaching of three bags of cocaine by an unknown third party to the outside of the vessel's hull constituted a malicious act. The Supreme Court held that "what the drafters appear to have in mind are persons whose actions are aimed at causing loss of or damage to the vessel, or, it may well be, other property or persons as a by-product of which the vessel is lost or damaged". Following the Court of Appeal decisions of The *Mandarin Star*[34] and The *Salem*,[35] Lord Mance held that in order to prove a malicious act, the assured

---

28 Danish Maritime Authority, *Analysis of Regulatory Barriers to the Use of Autonomous Ships: Final Report* (Danish Maritime Authority, December 2017).

29 Kevin Tester, "Technology in Shipping, the impact of technological change on the shipping industry" (*IMarEST and Clyde & Co*, November 2017) www.imarest.org/policy-news/thought-leadership/1010-technology-in-shipping/fil (accessed 30 December 2020).

30 UK P&I Club, "Q&A, Cyber risks and P&I insurance" (*Safety4Sea*, March 2018) https://safety4sea.com/wp-content/uploads/2018/03/UK-PI-Club-Cyber-Risks-and-PI-insurance-2018_03.pdf (accessed 30 December 2020).

31 Takehiko Hino, "Cyber risk and Cyber Security Countermeasures" (May 2018) 48 Loss Prevention Bulletin www.piclub.or.jp/wp-content/uploads/2018/05/Loss-Prevention-Bulletin-Vol.42-Full.pdf (accessed 30 December 2020).

32 International Underwriting Association, "Institute Cyber Attack Exclusion Clause" (*IUA*, 10 November 2003) www.iua.co.uk/IUA_Member/Clauses/IUA_Member/Clauses/eLibrary/Clauses.aspx?hkey=6f7dd1a3-6ab3-4b10-94c2-5a8c644b1c32 (accessed 20 December 2020).

33 *Navigators Insurance Co Ltd and others v Atlasnavios – Navegação LDA* (The *B Atlantic*) [2018] UKSC 26; [2019] AC 136.

34 [1968] 1 WLR 1325 (QB).

35 [1982] QB 946 (CA).

needs to prove that the act was done with ill will or spite towards the insured property, or the owner of other properties.

In reaching its decision, the Supreme Court did not follow the decision in The *Grecia Express*[36] in which Mr Justice Colman held that the words "persons acting maliciously" should be given a broader interpretation and, in light of the meaning under the Malicious Damage Act 1861, the words should "cover casual or random vandalism and do not require proof that the person concerned had the purpose of injuring the assured or even knew the identity of the assured". In *The B ATLANTIC*, the Supreme Court held that the reference to Malicious Damage Act 1861 was not helpful in the present case, since this Act could not have been in the minds of the drafters of the Institute Clauses in 1983.

The meaning of malicious acts in the context of insurance policies was also considered in *Tektrol v International Insurance of Hanover*,[37] a non-marine case which dealt with a contract of insurance for loss of a source code. The code was stored (i) in soft copy on a computer at a remote site; and (ii) in hardcopy in a pilot case at Tektrol's premises. The soft copy was erased after an employee accidently opened an email containing a piece of malware. The hard copy was subsequently stolen. The insurer denied coverage on the grounds that the loss of the source code from the remote site was caused deliberately by malicious persons (which was an excluded risk under the insurance policy). The first instance court held that the author who created the virus was a malicious person acting deliberately, even though his target was not directly on the assured's source code. On appeal, the assured argued that, to read the exclusion clause as a whole, the persons listed in exclusion clause as causing the damage were "rioters strikers locked-out workers persons taking part in labour disturbances or civil commotion or malicious persons", which clearly indicates that "the draftsman is speaking of interferences directed specifically at those computers and committed on or near the insured's premises". Therefore, if a remote hacker was included in the exclusion clause, it would introduce a completely different kind of attack. The Court of Appeal unanimously agreed with the assured's argument, and held that "the author of the virus was a 'malicious person', the clause does not extend to interferences by such people that are not directed at the computer systems, etc, used by the insured at the premis*es*".[38] In light of the above, it is difficult to reconcile the judgments in *The B ATLANTIC* and *Tektrol*.

Looking forward, if the doctrine in *The B ATLANTIC* were to be applied to future claims arising out of a cyber-attack, then it is likely that, unless the attack was specifically targeted at the assured then, the loss is unlikely to be excluded. Whilst some cyber-attacks are targeted in nature, such as spear phishing (where an email or some other electronic communication is sent, targeting a specific organisation or individual, seeking unauthorised access to sensitive information or funds), the majority of attacks are not (for example, such pieces of ransomware as NotPetya or WannaCry). Conversely, losses arising out of acts by disgruntled employees, such as the deliberate infection of computer systems, will be excluded.

There has, however, been increasing appetite in the market for an insurance product which provides cover for cyber risks. This has given rise to Clause 380 buy backs, by which

36 [2002] EWHC 203 (Comm Ct); [2002] 2 All ER 213 (Comm Ct).

37 *Tektrol Ltd v International Insurance Co of Hanover Ltd* [2005] EWCA Civ 845; [2006] 1 All ER 780 (Comm Ct).

38 *Tektrol Ltd v International Insurance Co of Hanover Ltd* [2005] EWCA Civ 845 [12];[2006] 1 All ER 780 (Comm Ct).

an insured, in return for an additional premium, can have Clause 380 removed from his hull insurance policy, and the creation of a number of bespoke cyber risk insurance policies for the marine market. The coverage of these policies varies enormously, from only providing business interruption cover to coverage for both hull and P&I risks.[39]

Most P&I Clubs do not exclude liability for P&I losses arising out of cyber risks. Similarly, the International Group Pooling Agreement is also not subject to a cyber risk exclusion. Members are, however, nevertheless obliged to ensure that cover is not prejudiced by acting in an "imprudent, unsafe, unduly hazardous or improper" way as doing so, depending on the rules of the P&I Club involved, may result in losses not being covered. This could include, for example, failing to carry out a cyber risk assessment and implement proper cyber security arrangements.

A shipowner may also not be indemnified for P&I losses due to a cyber risk where the losses were caused by war or an act of terrorism as war risks are normally specifically excluded from P&I cover. This should be particularly concerning for shipowners as the risk of war being conducted in cyberspace is seen as being increasingly likely. It is possible that, in the not too distant future, acts of war and/or terrorism will be conducted using cyber-attacks. Indeed, there have been a number of well-known instances where this may have already happened. First, in 2010, it was alleged that the first ever cyber weapon had been deployed by the United States and Israel against the Iranian nuclear program, specifically targeting the supervisory control and data acquisition ("SCADA") systems controlling the gas centrifuges used to separate nuclear material.[40] More specifically, stuxnet caused the fast-spinning centrifuges to tear themselves apart, reportedly ruining almost one-fifth of Iran's nuclear centrifuges after having infected 200,000 computers. Worryingly, stuxnet's design and architecture are not domain specific and could, theoretically, be tailored as a platform for attacking other types of SCADA and program logic controllers.

Turning to the issue of terrorism, whether or not a cyber-attack is an act of "terrorism" depends on the motivations of the author of the virus released or the hacker attacking systems. The Terrorism Act 2000[41] defines terrorism as being where the acts or threats are "made for the purpose of advancing a political, religious, racial or ideological cause".

The problem, however, for both shipowners and insurers in deciding whether insurance will respond following a loss is attribution. Save for a few well-known exceptions, the identities of the perpetrators behind the majority of attacks are unknown. Even if the geographical origin of an attack can be determined, it is very difficult to determine whether the attack was state sponsored or carried out by an individual, either politically motivated or even out of boredom.

Furthermore, shipowners should note that, even if a loss does not fall within the war or terrorism exclusion under a P&I wording, they may still not be able to recover the costs of any ransom that may have been paid to recover stolen data or unlock servers and so on. This is because many P&I Clubs either exclude the costs of ransom payments or leave it to the discretion of the Club's Board under the Omnibus Rule. In such cases, the Board will

39 For a more comprehensive analysis on these issues, see, B. Soyer, "Cyber-risk Insurance: Developing A New Cover in the Market" published in *Ship Operations: New Risks, Liabilities and Technologies in the Maritime Sector* (Informa Law, 2021), at 111–124.

40 K. Zetter, "An Unprecedented Look at Stuxnet, the World's First Digital Weapon" (*Wired*, 3 November 2014) www.wired.com/2014/11/countdown-to-zero-day-stuxnet/ (accessed 31 December 2020).

41 Terrorism Act 2000 (UK).

have a broad discretion and are likely to take into account whether payment of the ransom prevented losses which would have been insured in any event (e.g. injury to crew members or pollution).

If losses arising out of a cyber-attack are determined to have been an act of war or terrorism, then a shipowner would, in the first instance, look to its war risks insurance. However, if cover has been taken out on the Institute terms then the Institute Cyber Attack Exclusion CL380 will apply. The shipowner will therefore face the same issues addressed above in relation to hull and machinery insurance regarding whether or not the act was malicious. Conversely, other policy wordings take a different approach. For example, cover from the Hellenic War Risks Association currently excludes cover for losses caused by the use, as a means of inflicting harm, of any computer virus. However, the exclusion only applies once claims that would otherwise be excluded by the exclusion, exceed US$150 million across Hellenic's membership in any one policy year.[42]

Returning to the issue of seaworthiness, the implied warranty of seaworthiness is one of the most important parts of the Marine Insurance Act 1906 and, if the warranty is breached, the insurer may be relieved from liability. Section 39(1) of the Marine Insurance Act 1906 ("MIA") provides:

> In a voyage policy, there is an implied warranty that at the commencement of the voyage, the ship shall be seaworthy for the purpose of the particular adventure insured.

Conversely, section 39(5) of the MIA provides:

> In a time policy there is no implied warranty that the ship shall be seaworthy at any stage of the adventure, but where, with the privity of the assured, the ship is sent to sea in an unseaworthy state, the insurer is not liable for any loss attributable to unseaworthiness.

Section 39(4) MIA provides that a "ship is deemed to be seaworthy when she is reasonably fit in all respects to encounter the ordinary perils of the seas of the adventure insured". Before the advent of the Insurance Act 2015, breach of an implied warranty meant that the insurer was discharged of their liability to indemnify the owner of the vessel from the date of the breach (as per the general rule under the Marine Insurance Act 1906). Under this rule, there was no requirement to prove a causal link between the breach and the loss at all.

However, following the introduction of section 10 of the Insurance Act 2015, the situation has changed. Now, unless the parties have explicitly contracted out of this provision, the insurer's liability is automatically suspended, instead of being discharged, from the moment of breach of the implied warranty until the breach is remedied.

The Insurance Act 2015 does not, however, specify what is meant by the term "remedied". Section 10(5)(a) of the Act provides that a breach will be remedied if "the risk to which the warranty relates later become essentially the same as that originally contemplated by the parties". As the Insurance Act 2015 is relatively new, the Courts have not had an opportunity to clarify what is meant by "essentially the same".[43] Further, we expect that this issue will be very fact specific. In the context of MASS and cyber security, it could

42 M. Davey, J. Davey and O. Caplin, *Miller's Marine War Risks* (4th edn, Informa Law from Routledge, 2020), at para 5.14.

43 On this issue, see, B. Soyer, "Risk Control Clauses in Insurance Law: Law Reform and the Future" Cambridge Law Journal 75 (2016) 109, at 113–115.

mean that the warranty of seaworthiness would be remedied following, for example, the installation of a patch or software update.

Section 11 of the Insurance Act 2015 also gives rise to uncertainty in the context of autonomous vessels. The section provides:

> (1) This section applies to a term (express or implied) of a contract of insurance, other than a term defining the risk as a whole, if compliance with it would tend to reduce the risk of one or more of the following –
> (a) loss of a particular kind,
> (b) loss at a particular location,
> (c) loss at a particular time.
> (2) If a loss occurs, and the term has not been complied with, the insurer may not rely on the non-compliance to exclude, limit or discharge its liability under the contract for the loss if the insured satisfies subsection (3).
> (3) The insured satisfies this subsection if it shows that the non-compliance with the term could not have increased the risk of the loss which actually occurred in the circumstances in which it occurred.

It is currently unclear whether breach of the warranty of seaworthiness by the assured can be exempted. Section 11 provides that the assured will be indemnified if the assured can prove that the non-compliance with a warranty (or term) which tends to reduce the risk of loss of a particular kind, loss at a particular location or loss at a particular time could not have increased the risk of the loss which actually occurred in the circumstances in which it occurred. However, if a warranty or term intends to define the risk as a whole, then the aforementioned rules do not apply.

In considering this issue, the Law Commission has stated: "Our recommendation could apply to terms including warranties (including the implied marine warranties)".[44] It therefore appears that, in the eyes of the Law Commission, the implied warranty of seaworthiness may not be a term "defining the risk as a whole". This is surprising given the central importance of seaworthiness to shipping. The authors of Arnould are, however, firmly of the view that s. 11 of the Insurance Act 2015 does not apply to the warranty of seaworthiness, based on the following: first, the target of the warranty of seaworthiness is not to reduce the risk of such loss; second, perils of the sea are a class of perils of indefinite extent rather than a particular kind.[45] This discussion is important in the context of autonomous vessels because, if there was a loss arising out of a failure to comply with domestic or international law in relation to cyber security, then in the absence of an express term covering cyber security, the only term available to an insurer to avoid liability will be the warranty of seaworthiness. Further, if the policy did include express terms requiring compliance with a certain level of cyber security, then it is arguable that such a term can be viewed as a term

44 Law Commission, *Insurance Contract law: Business Disclosure; Warranties; Insurers' Remedies for Fraudulent Claims; and Late Payment* (Cm 8898, Law Com No. 353, 2014), at para 18.41.

45 J. Gilman, M. Templeman, C. Blanchard, P. Hopkins and N. Hart, *Arnould: Law of Marine Insurance and Average* (19th edn, Sweet & Maxwell, 2019), at para 20–08. For a similar view, see, B. Soyer, "Insuring Cargoes in the New Era: Impact of the Insurance Act 2015 on Standard cargo Clauses/Wordings" in *International Trade and Carriage of Goods* (Informa Law, 2017), 264, at 267–269.

that tends to reduce the risk of loss of a particular kind, meaning that the term would fall within the scope of Section 11 of the Insurance Act 2015.

5.4 Conclusion

Despite the fact that, technologically speaking, shipping is entering a new world of innovation, much of the existing legal regime is equally applicable to MASS and conventional vessels. Nevertheless, there are a number of questions which have not been dealt with by the Courts yet and these are likely to be the subject of judicial interpretation in future.

CHAPTER 6

Use of unmanned aircraft systems in a maritime context

Operational, regulatory and legal issues

George Leloudas and Michael Chatzipanagiotis***

6.1 Unmanned aircraft systems

Unmanned Aircraft Systems (UAS), commonly known as "drones", can be defined as arrangements for an aircraft to operate autonomously or to be flown without a pilot on board.[1] An unmanned aircraft *system* consists of the unmanned aerial vehicle itself (the UAV), a remote control station and a datalink between the UAV and the control station.

The main operating method of a UAS is within the Visual Line of Sight (VLOS) of the remote pilot who handles the control station. The pilot is required to retain visual access to the UAV without the use of visual enhancements, such as on-board cameras, binoculars, telescopes, or visual observers.[2] The UK Civil Aviation Authority (UK CAA) presumes that VLOS requirements are satisfied when the UAV is flown out "to a distance of 500 metres horizontally from the remote pilot".[3]

An alternative method of operation is Beyond the Visual Line of Sight (BVLOS) of the remote pilot. BVLOS operations do not require the remote pilot to maintain visual contact with the UAV during flight, enabling its operation at distances longer than 500 m from the pilot. While remote pilots in VLOS operations must comply with visual flight rules and have primary responsibility to avoid collisions,[4] a UAS in BVLOS mode must be able to perform an equivalent function in terms of detecting and avoiding collisions ("Detect-and-avoid", or DAA).[5] Establishing a DAA capability is a fundamental requirement for the UK CAA to grant permission to operate BVLOS.[6]

* Associate Professor, Member of the Institute of International Shipping and Trade Law (IISTL), Swansea University.

** Lecturer in Private Law, University of Cyprus.

1 See e.g. Commission Delegated Regulation (EU) 2019/945 of 12 March 2019 on unmanned aircraft systems and on third-country operators of unmanned aircraft systems, C/2019/1821, OJ L 152, 11.6.2019, at 1–40, art. 2(1) and Air Navigation Order 2016, SI 2016/765.

2 See the UK CAA, "An Introduction to Unmanned Aircraft Systems" www.caa.co.uk/Consumers/Unmanned-aircraft/Our-role/An-introduction-to-unmanned-aircraft-systems/ (accessed 12 February 2021).

3 UK CAA, *CAP 722: Unmanned Aircraft System Operations in UK Airspace – Guidance* (CAP 722, 2020), para 2.1.1 (CAP 722).

4 Ibid., paras 3.6.1 and 3.6.4.

5 UK CAA, "Beyond Visual Line of Sight in Non-Segregated Airspace Fundamental Principles & Terminology" (October 2020) https://publicapps.caa.co.uk/docs/33/CAP%201861%20-%20BVLOS%20Fundamentals%20v2.pdf (accessed 12 February 2021).

6 See below, 6.3.3.

DOI: 10.4324/9781003155195-6

A third method of operation is the Extended Visual Line of Sight (EVLOS). This is a subcategory of BVLOS operations but requires the grant of a relevant operating permission. Under EVLOS operations, the DAA function is performed by visual observers who relay information to the remote pilot about the operation of the UAV. The observers are required to have continuous, unaided, visual contact with it.[7]

The Inertial Measurement Unit (IMU) is one of the most important components of a UAV. By using several gyroscopes, accelerometers and magnetometers, the IMU controls the horizontal and vertical movements of the UAV, ensuring the smooth flying that is crucial for taking high-quality images and creating 3D maps.[8] The data collected by the IMU are fed to the flight control unit, which signals the speed controllers to regulate the motors and the propellers in terms of thrust and speed.[9]

What has increased the appeal of UASs for commercial users is their GPS enablement and their ability to carry a wide range of sensors. The use of GPS antennas enables the UAV to communicate its position to the remote pilot with accuracy, to maintain position at a fixed location and avoid going above the permissible altitude or inside "no-fly" zones. GPS antennas also enable the UAV to return to the take-off spot under prescribed circumstances, such as when the datalink is lost or the battery level falls below a prescribed level, to identify a landing location in emergency situations and, where permissible, to operate in BVLOS mode.[10]

Furthermore, manufacturers and operators of UASs increasingly install vision systems on them, such as high-definition cameras and LED lighting systems, infrared and ultrasonic distance sensors, as well as photogrammetry and Light Detection and Ranging (LiDAR) sensors. These in turn enable UASs to be connected to cellular and/or wireless networks, so improving the communication links with them. The combination of these sensors enhances their operational reliability and the quality of data collection for the following reasons: i) they can sense and avoid obstacles by flying around them or hovering in front of them; ii) they can operate in areas or spaces where there is no GPS signal, or magnetic interference affects the operation of the internal compass;[11] iii) they can hover perfectly still, a function that improves the quality of any images taken; iv) they can broadcast live videos and real-time flying information to distant control centres;[12] and v) with the increased use of LiDAR sensors, they can collect detailed data to conduct "material volume and tonnage calculations, and . . . floorplan and building measurements".[13]

---

7 CAP 722, see no. 3, para 2.1.2.1.

8 F. Corrigan, "Drone Gyro Stabilization, IMU And Flight Controllers Explained" (DroneZon, 7 May 2020) www.dronezon.com/learn-about-drones-quadcopters/three-and-six-axis-gyro-stabilized-drones/ (accessed 12 February 2021).

9 Ibid.

10 DroneDeploy, "Safety Considerations and Avoiding Flyaways" at https://support.dronedeploy.com/docs/working-safety-considerations#:~:text=GPS%20is%20used%20by%20your,know%20its%20position%20in%20space.&text=The%20compass%20allows%20the%20drone,heading%20or%20direction%20in%20space (accessed 12 February 2021).

11 Z. Dukowitz, "What Are GPS-Denied Drones and Why Are They Important?" (*UAV Coach*, 2020) https://uavcoach.com/gps-denied-drones/ (accessed 12 February 2021).

12 Ericsson, "How mobile networks can support drone communication" www.ericsson.com/en/blog/2017/11/how-mobile-networks-can-support-drone-communication (accessed 12 February 2021).

13 Heliguy, "Lidar Sensors" www.heliguy.com/lidar-sensors-i174 (accessed 12 February 2021); NASA, "Autonomous Drone Navigation System Ends Reliance on GPS" (2020) https://spinoff.nasa.gov/Spinoff2020/ps_5.html (accessed 12 February 2021).

UAVs used in maritime operations are usually light. Their maximum take-off mass (MTOM) is generally less than 25 kg, and they have four or six rotary wings. It is common for maritime operators to use off-the-shelf UASs; however, they often modify them by fitting protective shields and a variety of additional sensors and cameras. At the same time, custom-made UASs, such as the Elios 2 of Flyability and the Custom Drone of DNVGL, have gained popularity as they are designed to cater for the special needs of maritime environments, namely operating without GPS signal, flying in BVLOS mode and/or being used in places which regularly experience "turbulences due to the small air volume or to drafts . . . presence of dust . . . complete darkness and presence of reflective surfaces".[14]

The aim of this chapter is to provide, for the first time, a comprehensive analysis of a wide range of regulatory, insurance and liability issues that are raised by using UASs in a commercial maritime context. It examines how special characteristics of operating UASs in the maritime industry impact their licensing at a European and UK level. It also investigates the liability regime for property damage/personal injury caused by the operation of UASs in the UK, and evaluates whether current regulatory insurance requirements, standard UAS insurance policies and insurance cover provided by hull insurers and P&I Clubs are sufficient to deal with the liability challenges stemming from the increased use of UAS. Our analysis does not deal with privacy-related issues of UASs as these do not, currently, raise significant concerns in the maritime industry.

6.2 Commercial maritime uses of unmanned aircraft systems

Classification societies and operators of offshore installations are prominent among the maritime users of UASs.

Classification societies increasingly use UASs to inspect cargo and ballast tanks, as well as structures on deck and hull, for damage. They also use them to undertake inspections at high areas, replacing scaffolding and/or cherry pickers, and also inside confined spaces to reduce the exposure of workers to noxious gases.[15] Currently, UASs are not deployed in tanks containing explosive atmospheres;[16] however, some "explosion proof" UASs have been certified for use.[17] In November 2020, the Korean Register of Shipping undertook its first UAS survey of a bulk carrier without using any scaffolding. The survey was undertaken by a UAS that inspected the higher places and a crawler, namely a drone that can climb up walls, that measured the thickness of the hull.[18]

Similar benefits are experienced by operators of offshore installations, as UAS inspections provide alternatives to "more traditional, time consuming and costly inspection methods like rope access, sky-lifts, cherry pickers and scaffolding".[19] Furthermore, UASs

14 Flyability, "Confined Spaces Inspection" at www.flyability.com/articles-and-media/confined-spaces-inspection (accessed 12 February 2021).

15 See ClassNK, "Guidelines for Use of Drones in Class Surveys" (April 2018).

16 Flyability, "Are Drones Intrinsically Safe?" at www.flyability.com/articles-and-media/are-drones-intrinsically-safe (accessed 18 February 2021).

17 E. Knutt, "Indoor Explorers. Lone Workers" (2018) Tolley's Health and Safety at Work, 7: 32, 33.

18 C. Jallal, "KR Completes First Drone and Crawler Hull Inspection Survey" (Riviera, 16 November 2020) www.rivieramm.com/news-content-hub/kr-completes-first-drone-and-crawler-hull-inspection-survey-61808 (accessed 18 February 2021).

19 Airborne Drones, "Offshore Oilrig Inspections with Drones" (23 April 2019) www.airbornedrones.co/offshore-oilrig/ (accessed 18 February 2021).

enable operators to undertake internal and external flare stack inspections without shutting the entire installation down.[20] As early as February 2016, the flare system of the Berkut oil rig, the biggest rig in the world, was inspected by a UAS under weather conditions which limited the performance of its batteries to 20 minutes and prevented the receiving of GPS signal.[21] In addition, in June 2020, a UAS inspected the oil tank of a floating production storage and offloading (FPSO) vessel for the first time.[22] As no GPS signal was available inside the tank, LiDAR sensors were used to create "a 3-D map of the tank and all images and video [were] accurately geo-tagged with position data".[23]

The increasing popularity of UASs among vessel and offshore operators stems from their potential to minimise casualties and reduce the cost of operations, while increasing their efficiency. UASs enable inspection and maintenance tasks to be monitored remotely on a real-time basis, and can take high-resolution photos that reduce the time necessary to take remedial action.[24] At the same time, costs are reduced as UAS surveys do not require the vessel or the oil rig to be placed out of operation for several weeks.[25] It has been estimated that their use in the offshore sector creates savings of $US 1 million per day.[26]

Furthermore, UASs are tested to be used to transfer mooring lines between ships and tugs or quaysides,[27] to take photos of salvage projects before salvors commit resources in salvage operations,[28] and to provide video for security reasons, especially when navigating in piracy-infested waters.[29]

To improve the efficiency of everyday operations, UASs have also been tested as a means of transferring spare parts[30] and 3D-printed consumables, as well as documents and crew supplies, to and from vessels.[31] In November 2020, the first night delivery took place at the port of Singapore: a 3D-printed part weighing 3 kg was carried by a UAS to the *Berge*

20 Ibid.

21 R. Knight, "UAV Inspection at the Biggest Oil Rig in the World" (*Microdrones*, 2 December 2019) www.microdrones.com/en/content/uav-inspection-at-the-biggest-oil-rig-in-the-world/ (accessed 18 February 2021).

22 UNCTAD, "Review of Maritime Transport 2020" (United Nations, 2020), at 123.

23 C. O'Reilly, "Autonomous Drone Inspections Move Step Closer After Successful Test" (*Hydrocarbon Engineering*, 9 June 2020) www.hydrocarbonengineering.com/tanks-terminals/09062020/autonomous-drone-inspections-move-step-closer-after-successful-test/ (accessed 18 February 2021).

24 M. Wingrove, "Drones Raise Accuracy and Safety of Ship Surveys" (*Riviera*, 27 October 2020) www.rivieramm.com/news-content-hub/drones-raise-accuracy-and-safety-of-ship-surveys-61446 (accessed 18 February 2021).

25 Airborne Drones, see no. 19.

26 J. Karpowicz, "4 Ways Drones Are Being Used in Maritime and Offshore Services, (*Commercial UAV News*, 9 August 2018) www.commercialuavnews.com/security/4-ways-drones-maritime-offshore-services (accessed 18 February 2021).

27 M. Wingrove, "Five Technologies to Change Tug Operations Forever" (*Riviera*, 18 January 2018) www.rivieramm.com/news-content-hub/news-content-hub/five-technologies-to-change-tug-operations-forever-25968 (accessed 18 February 2021).

28 M. Wingrove, "5 Technologies to Transform Salvage" (*Riviera*, 29 December 2020) www.rivieramm.com/news-content-hub/technologies-to-transform-salvage-62178 (accessed 18 February 2021).

29 Airborne Drones, "Surveillance & Security Drone" at www.airbornedrones.co/surveillance-and-security/ (accessed 18 February 2021).

30 Safety4Sea, "Drones to Deliver 3D Printed Spare Parts in Singapore" (11 June 2020) https://safety4sea.com/drones-to-deliver-3d-printed-spare-parts-in-singapore/#:~:text=Wilhelmsen's%20Marine%20Products%20division%20has,over%20a%20distance%20of%2050km.&text=F%2Ddrones%20now%20aims%20to,vessels%20up%20to%20100km%20away (accessed 18 February 2021).

31 H. McNabb, "Drone Delivery to Ships Takes off in Singapore" (*DroneLife*, 29 April 2020) https://dronelife.com/2020/04/29/drone-delivery-to-ships/#:~:text=F%2Ddrones%20completed%20the%20first,%2Ddrones'%20first%20paying%20customer (accessed 18 February 2021).

Sarstein that was anchored about three nautical miles outside the port of Singapore, with the flight completed in seven minutes.[32] It is expected that a UAV that can carry 100 kg over 100 km will be tested before the end of 2021, enabling, if successful, deliveries of heavy items to vessels en route.[33] In this context, UASs have the potential to reduce the demand for supply vessels, and can thus incidentally contribute to the reduction of CO_2 emissions.[34]

In the long term, UASs are expected to play a central role in the digitisation of shipping operations. Digitisation is expected to reach new heights with the introduction of the "digital twin" concept, where the physical performance of a vessel will be represented digitally from its construction and throughout its lifetime. Real-time performance data will be stored in cloud-based systems that can be shared with manufacturers and/or regulators, and will make recommendations for optimising the vessel's performance.[35] It is estimated that UASs will be one of the means of receiving real-time data from vessels to accurately predict their voyage performance in terms of fuel efficiency and emissions, to evaluate the condition of the hull and machinery, to schedule predictive maintenance, and to reduce accidents in busy ports and shipping lanes by providing forward navigation information.[36]

Currently, UASs are part of a significant investment by the maritime industry in remote inspection technologies that also include "unmanned robot arm[s], remotely operated vehicles, climbers".[37] The restrictions imposed by COVID-19 have accelerated the frequency of their use, yet their impact was felt well before the pandemic:

> DNV GL . . . has reportedly completed 15,000 remote surveys since their launch in October 2018. One in five of Lloyd's Register surveys are completed remotely. In March 2020, the number of complex remote surveys performed by LR increased by 25 percent. RINA began trialling remote inspections in May 2019 and has completed approximately 300 inspections between May 2019 and February 2020, and another 60 inspections in March 2020.[38]

Reflecting their increased use, several classification societies and the International Association of Classification Societies (IACS) have published guidelines on the use of remote inspection techniques.[39] A review of them reveals that most societies use external providers to perform UAS surveys. For example, the guidelines of ClassNK refer to "drone service suppliers" which are defined as companies that are not employed by the society, but provide "survey, inspection and similar services for ships . . . by using drones, based on an application from the manufacturer, shipyard, ship owner or other client".[40] While societies use the data

32 F-Drones, "World's First Commercial Night Drone Delivery by F-Drones" (4 November 2020) www.f-drones.com/post/world-s-first-commercial-night-drone-delivery-by-f-drones (accessed 18 February 2021).

33 UNCTAD, see no. 22, at 123.

34 H. McNabb, see no. 31.

35 See DNV GL, "Digital Twins for Blue Denmark" (2 January 2018) www.dma.dk/Documents/Publikationer/Digital%20Twin%20report%20for%20DMA.PDF (accessed 18 February 2021).

36 Riviera, "Class is Adopting Drone Technology" (15 August 2016) www.rivieramm.com/opinion/opinion/class-is-adopting-drone-technology-1-53170 (accessed 18 February 2021).

37 Bureau Veritas, "Approval of Service Suppliers" (January 2020), at para 14.1.2. http://erules.veristar.com/dy/data/bv/pdf/533-NR_2020-01.pdf (accessed 18 February 2021).

38 SeaDrone, "Class Societies' Steady March to Remote Inspection Technologies and Techniques" (24 August 2020) https://seadronepro.com/blog/class-societies-steady-march-to-remote-inspection-technologies-and-techniques (accessed 18 February 2021).

39 For example, the ClassNK, the American Bureau of Shipping, Lloyd's Register and Burau Veritas have published best practices on using UAS during inspections.

40 ClassNK Guidelines, *supra* no. 15, [1.4]. Similar wordings are used in the "Guidance Notes on the Use of Remote Inspection Techniques" of the American Bureau of Shipping and the "Approval of Service Suppliers" of Bureau Veritas.

collected by those surveys to further develop remote inspection techniques,[41] their preferred practice is to outsource the risk and the cost of research and development to UAS operators.

At the same time, a significant concern is the regulatory and liability framework of UAS operations. It has been argued that the legal and regulatory background is in a state of flux and slows down their further deployment especially with respect to BVLOS operations. Often the gaps are filled with tailor-made solutions, such as the one proposed by the American Bureau of Shipping which recommends the use of the regulatory framework of the FAA or the UK CAA where no national UAS requirements are in force.[42] Although a time-lag between the deployment of a new technology and the creation of a comprehensive regulatory and liability regime is to be expected, our work examines whether the current systems deserve such criticism. In this context, the next section examines the regulatory challenges of licensing commercial UASs in Europe and the UK.

6.3 Regulatory requirements for commercial unmanned aircraft systems

There is a distinction among the regulatory requirements for UAS imposed by international law, EU law and UK law.

6.3.1 International law

In international law, rules applicable to UASs can be found in the International Convention on Civil Aviation (the Chicago Convention – hereafter the CC)[43] and in the UN Convention on the Law of the Sea (UNCLOS).[44]

The Chicago Convention applies only to civil aircraft, not to state aircraft used mainly in military, customs and police services.[45] Art. 8 of the CC prohibits flights of pilotless aircraft over the territory of a contracting State without special authorisation from that State, while contracting States for their part undertake to ensure the safety of such flights as to the risks posed to other airspace users. The territory of a State includes the land areas and its territorial waters.[46] Ports are deemed part of the coast.[47] Moreover, Art. 94 of UNCLOS provides for the jurisdiction of the flag State over ships flying its flag over the High Seas. The combined result of these provisions is that the territorial State and, over the High Seas, the flag State have jurisdiction as to the use of UASs.

At the same time, the CC makes also clear that UAS are "aircraft" and thus are, in principle, subject to all provisions applicable to manned aircraft. Therefore, UASs are to be registered as aircraft, which provides them with their own nationality and subjects them to the jurisdiction of the State of registry.[48] Hence, there may be more than one

41 Ibid.

42 American Bureau of Shipping, "Guidance Notes on the Use of Remote Inspection Technologies" (February 2019), [3].

43 Convention on International Civil Aviation, done at Chicago on 7 December 1944, in force since 4 April 1947, 15 UNTS 295.

44 United Nations Convention on the Law of the Sea, done at Montego Bay on 10 December 1982, in force since 16 November 1994, 1833 UNTS 3.

45 CC, art. 3(a)-(b).

46 CC, art. 2, UNCLOS, art. 2(2).

47 UNCLOS, art. 11.

48 CC, art. 17.

State competent in respect of incidents connected to the maritime uses of UASs. As a rule, territorial jurisdiction prevails.[49] However, in practice, each State is concerned with issues affecting its interests, which means that different aspects of UAS operations may interest different States. For instance, a coastal State is likely to be uninterested in occupational accidents that occur in its territorial waters and involve a foreign ship and a foreign UAS, to the extent that the repercussions of the accident are confined strictly to the vessel itself and do not affect navigation. Nevertheless, from the viewpoint of ship masters and UAS operators, compliance with the regulations of all States involved is advisable.

Furthermore, UASs are subject to the technical rules contained in the Annexes to the CC, developed under Art. 38.[50] Nonetheless, there are particularities of UASs, which vary depending on their specifications and their mission profile, and render necessary the adjustment of the rules on manned aircraft. The UAS-specific regulatory activities of the International Civil Aviation Organisation (ICAO) have so far been limited to advisory materials and regulatory support to its member States on developing national legislation. The ICAO has developed model UAS legislation and related Advisory Circulars,[51] and has established a special advisory group on UASs.[52]

Of particular importance for maritime UAS operations is Annex 18 on *The Safe Transport of Dangerous Goods by Air* and its implementing rules and procedures.[53] Especially regarding UAS operations, the ICAO has issued Advisory Circular 102–37, which prescribes the duty of aircraft operators to establish Standard Operation Procedures for carriage of dangerous goods, which must include at least (a) how to conduct a safety risk assessment, (b) a personnel training programme to ensure competency, (c) instructions to be communicated to persons related to dangerous goods in case of an incident or accident and (d) instructions for the collection of safety data.[54]

6.3.2 EU law

In the EU, civil UAS operations are regulated mainly by the Implementing Regulation (EU) 2019/947[55] and the Delegated Regulation (EU) 2019/945,[56] which specify the

49 See CC, art. 1 which provides for complete and exclusive sovereignty of the territorial State over the airspace above its territory.

50 For a list of the Annexes see ICAO e-library at https://elibrary.icao.int/home (accessed 18 February 2021).

51 ICAO, "Introduction to ICAO Model UAS Regulations and Advisory Circulars" at www.icao.int/safety/UA/Pages/ICAO-Model-UAS-Regulations.aspx (accessed 18 February 2021).

52 ICAO, "Unmanned Aircraft Systems Advisory Group" at www.icao.int/safety/UA/Pages/Unmanned-Aircraft-Systems-Advisory-Group-(UAS-AG).aspx (accessed 18 February 2021).

53 ICAO, *Technical Instructions for the Safe Transport of Dangerous Goods by Air* (ICAO Doc 9284–2018).

54 ICAO, *Unmanned Aircraft Systems (UAS) Carrying Dangerous Goods* (Revised Advisory Circular 102–37).

55 Commission Implementing Regulation (EU) 2019/947 of 24 May 2019 on the rules and procedures for the operation of unmanned aircraft, C/2019/3824, OJ L 152, 11.6.2019, pp. 45–71 (UAS Implementing Regulation).

56 Commission Delegated Regulation (EU) 2019/945 of 12 March 2019 on unmanned aircraft systems and on third-country operators of unmanned aircraft systems, C/2019/1821, OJ L 152, 11.6.2019, at 1–40 (UAS Delegated Regulation).

requirements laid down in Arts 55–58 of the European Aviation Safety Agency (EASA) Basic Regulation.[57]

6.3.2.1 General principles

The general objective of the regulation of UAS operations in the EU is they should be as safe as those in manned aviation.[58]

Three main principles form the basis of UAS regulation in the EU. The rules are: (1) operation-centric, since the consequences of an accident or incident with a UAS are highly dependent on the environment where the accident or incident takes place; (2) risk-based, which means that they focus on the risk created by the operation; (3) performance-based, in that they identify the required level of performance, without prescribing exact means to achieve it.[59]

6.3.2.2 Regulatory categories

The Regulations adopt a risk-based approach and do not distinguish between leisure or commercial activities. Important are the technical specifications and the intended operations of the UAS. On this basis, the Regulations lay down three regulatory categories: (1) the Open Category, (2) the Specific Category and (3) the Certified Category.

6.3.2.2.1 Open category

The Open Category comprises UASs whose specifications and mission profile is considered to pose little risk. It is divided into three subcategories A1, A2 and A3.

Art. 4 of the UAS Implementing Regulation (EU) 2019/947 provides that the Open Category applies to UAS with an MTOM[60] of less than 25 kg, which will fly at a safe distance from assemblies of people.[61] Such UASs need to be operated within the VLOS of their operator, and at an altitude of less than 120 m from the ground. Carriage of dangerous goods,[62] or dropping of any material, is not allowed.

The Open Category is divided into three subcategories, A1, A2 and A3, depending on the operational limitations, requirements for remote pilots and technical requirements for

---

57 Regulation (EU) 2018/1139 of the European Parliament and of the Council of 4 July 2018 on common rules in the field of civil aviation and establishing a European Union Aviation Safety Agency, and amending Regulations (EC) No. 2111/2005, (EC) No. 1008/2008, (EU) No. 996/2010, (EU) No. 376/2014 and Directives 2014/30/EU and 2014/53/EU of the European Parliament and of the Council, and repealing Regulations (EC) No. 552/2004 and (EC) No. 216/2008 of the European Parliament and of the Council and Council Regulation (EEC) No. 3922/91, OJ L 212, 22.8.2018, at 1–122.

58 UAS Implementing Regulation, op. cit., no. 55, recital (3).

59 See EASA, "Opinion No. 01/2018. *Introduction of a Regulatory Framework for the Operation of Unmanned Aircraft Systems in the 'Open' and 'Specific' Categories*" at www.easa.europa.eu/sites/default/files/dfu/Opinion%20No%2001-2018.pdf (accessed 18 February 2021).

60 Under Art. 2(22) of UAS Implementing Regulation, op. cit., no. 55, the MTOM includes the mass of the unmanned aircraft, including payload and fuel, as defined by the manufacturer.

61 Art. 2(3) of UAS Implementing Regulation, op. cit., no. 55, defines "assemblies of people" as gatherings where persons are unable to move away due to the density of the people present.

62 Art. 2(11) of UAS Implementing Regulation, op. cit., no. 55, defines "dangerous goods" as articles or substances, which can pose a hazard to health, safety, property or the environment in the case of an incident or accident, that the unmanned aircraft is carrying as its payload, such as explosives, gases, flammable liquids, flammable solids, oxidising agents, toxic and infectious substances (poison, biohazard), radioactive substances and corrosive substances.

UAS.[63] In this regard, each UAS has to bear a class identification label, ranging from C0 to C4, which corresponds to product-safety requirements.[64]

Operators are not required to obtain an authorisation by the competent Civil Aviation Authority (CAA), nor to submit a special declaration to the CAA regarding the safety of the UAS operation.[65] Safety of operations is established through product-safety requirements[66] and through organisational duties imposed on operators, for example development of appropriate operational procedures, personnel training, confinement of operations into a specific area and so on.[67]

In the maritime context, the Open Category will comprise UAS operations of small aircraft, the operational profile of which cannot pose a risk to personnel on board or near the ship (for example, hull inspections and inspections of interior spaces with no persons in them).

6.3.2.2.2 Specific category

The Specific Category comprises UAS operations that do not meet the criteria of the Open Category. In these cases, the operator is required, in principle, to obtain an operational authorisation from the competent authority.[68]

To obtain the authorisation, the operator has to perform a risk assessment and to adopt appropriate measures to minimise risk.[69] Such assessment has to describe the characteristics of the UAS operation, identify potential risks to third parties or property and propose adequate mitigating measures.[70]

However, an authorisation is not required: (a) for an operation complying with a "Standard Scenario",[71] in which an operational declaration by the operator is sufficient; or (b) when the operator holds a Light UAS Operator Certificate with the appropriate privileges, granting it the ability to authorise its own operations, under specific terms and conditions, without the need of an operational declaration or a CAA authorisation.[72]

The Specific Category will apply to a large number of UASs used in maritime operations as it covers flights of light UAS, that is, with a MTOM less than 25 kg, conducted near the personnel of the vessel or other people.

6.3.2.2.3 Certified category

The Certified Category includes UASs involved in operations involving a high level of risk.

---

63 See in detail the Annex, Part A, to the UAS Implementing Regulation, op. cit., no. 55.

64 See in detail the Annex to the UAS Delegated Regulation, op. cit., no. 56. These requirements will become applicable as of 1 January 2023.

65 UAS Implementing Regulation, op. cit., no. 55, art. 3(a).

66 See, in detail, UAS Delegated Regulation, op, cit., no. 56, art. 2(1) and arts 4–39.

67 See details in Annex A of UAS Implementing Regulation, op. cit., no. 55.

68 UAS Implementing Regulation, op. cit., no. 55, arts 3 and 12.

69 UAS Implementing Regulation, op. cit., no. 55, art. 5(2).

70 See UAS Implementing Regulation, op. cit., no. 55, art. 11.

71 Art. 2(6) of UAS Implementing Regulation, op. cit., no. 55, defines a "standard scenario" as a type of UAS operation in the SC category, as defined in UAS Delegated Regulation, for which a precise list of mitigating measures has been identified in such a way that the competent authority can be satisfied with declarations in which operators declare that they will apply the mitigating measures when executing this type of operation.

72 See, in detail, Part C of the Annex to the UAS Implementing Regulation, op. cit., no. 55.

It applies to UASs which (a) have a characteristic dimension of at least 3 m and are designed to be operated over assemblies of people; (b) are designed for transporting people; (c) are designed for transporting dangerous goods; or (d) are considered by the competent authority to pose operational risks that can only be adequately mitigated through the certification process.[73]

The EASA observes that UASs in the Certified Category need to comply with the rules applicable to manned aircraft, such as the need for a type certificate and a certificate of airworthiness, which in practice requires vigorous and time-consuming efforts. The EASA plans to gradually develop such rules.[74]

UAS in maritime operations will be covered by the Certified Category mainly in cases of transport of dangerous goods to and from ships or overseas platforms, for example batteries, potentially infected blood samples and so on.

6.3.2.3 Registration

Art. 14 of the Implementing Regulation (EU) 2019/947 provides for the registration of both UAS and their operators.

The owner of a UAS that is subject to certification (Certified Category) needs to have it registered in accordance with Annex 17 to the CC.

UAS operators are required to register in the State of their residence for natural persons, or their principal place of business for legal entities. The registration duty applies to operators of UASs of the Specific Category. It also applies to operators of the Open Category, when operating an aircraft which (a) has an MTOM of at least 250 g or more, or (b) in the case of an impact can transfer to a human kinetic energy above 80 Joules, or (c) is equipped with a sensor able to capture personal data.

6.3.2.4 National implementing measures and enforcement

Although the Regulations are directly applicable in all EU Member States, implementing national legislation is required for issues not covered by the EU instruments. These are mainly enforcement issues[75] and domains left in the discretion of the Member States, such as operational conditions for UAS geographical zones for safety, security, privacy or environmental reasons.[76]

As to enforcement, EU rules on UAS are enforced by the competent national authorities, which are usually the Civil Aviation Authorities of each Member State.

6.3.3 UK law

The European Union (Withdrawal) Act 2018 repealed the European Communities Act 1972 and brought an end to the effect of European legislation in the UK.[77] Consequently,

73 UAS Delegated Regulation, op. cit., no. 56, art. 40(1) and UAS Implementing Regulation, op. cit., no. 55, art. 6(1).

74 See EASA, "Certified Category – Civil Drones" at www.easa.europa.eu/domains/civil-drones-rpas/certified-category-civil-drones (accessed 18 February 2021).

75 See e.g. UAS Delegated Regulation, op. cit., no. 56, arts 19–20 and UAS Implementing Regulation, op. cit., no. 55, art. 17–19.

76 See UAS Implementing Regulation, op. cit., no. 55, art. 15.

77 European Union (Withdrawal) Act 2018, s. 1.

a new body of UK law has been created: retained EU law. It contains, among other legal instruments, EU Regulations which are amended to operate appropriately in a UK legal context. Reflecting this change, the analysis in this chapter follows, in principle, the text of the UK Regulations as retained with references to the text of the European ones where required.

In the UK, UAS operations are governed mainly by retained EU law[78] and the Air Navigation Order 2016 (ANO).[79] Moreover, the UK CAA has issued detailed guidance regarding UAS operations in the UK airspace.[80]

A question to be determined is if and to what extent the UK will follow any EU rules that enter into force after 31 December 2020. It is likely that the UK will decide whether to do so on a case-by-case basis. For example, the UK CAA has stated that, since the EU requirements for a Standard Scenario and a Light UAS operator certificate will become applicable on 31 December 2020, UK operators should proceed under the pre-existing UK operational authorisation process.[81]

The ANO supplements the retained EU Regulations. Thus, it applies to endangerment regulations and legal penalties for breaches of these regulations,[82] and also specifies general requirements referred to in the retained EU Regulations, for example airspace restrictions around aerodromes and other locations.[83] Moreover, the ANO applies to details of operations of UASs of the Certified Category, since the EU has not yet developed detailed rules in this regard.

One of the most pressing demands of commercial operators in the UK is for the UK CAA to facilitate the operations of UASs in BVLOS mode that will enable their operation in far greater distances and from centres located away from the immediate vicinity of the UAS.[84] To mitigate the increased risk of collisions in BVLOS operations, the UK CAA grants permissions if one of the following three circumstances are satisfied: (1) a DAA capability has been integrated into the UAS; or (2) the UAS operates in segregated airspace, namely airspace that has been reserved for the operation of the UAS; or (3) the operator provides clear evidence that the UAS flight poses "no aviation threat" and that the safety of persons and objects on the ground has been properly addressed.

With respect to the required DAA capability of the UAS, the UK CAA has stressed that installing live-feed cameras on board the UAS is not an acceptable means of demonstrating that it has DAA capability. For the permission to be granted, the installed technology is required to take autonomous decisions by being able to "observe the environment surrounding the drone, decide whether a collision is imminent, and generate a new flight path in order to avoid collision".[85] Currently, such systems employ a variety of active sensors,

78 The UK has retained with amendments the UAS Implementing Regulation, op. cit., no. 55 and the UAS Delegated Regulation, op. cit., no. 56.

79 Air Navigation Order 2016, SI 2016/765.

80 CAP 722, op. cit., no. 3.

81 UK CAA, "The EU UAS Regulation Package" (CAP 1789, June 2020), at 10.

82 See e.g. ANO, op. cit., no. 79, ss. 240–241 and 257.

83 CAP 722, op. cit., no. 3, para. 1.2.4.

84 See, op. cit., Chapter I.

85 Unmanned Systems Technology, "Sense and Avoid Technology" at www.unmannedsystemstechnology.com/category/supplier-directory/electronic-systems/sense-avoid-systems/#:~:text=Sense%20and%20Avoid%20(SAA)%20or,lines%2C%20birds%20and%20other%20obstacles (accessed 18 February 2021) and A. Galindo, "Detect-and-Avoid (DAA): How Airborne Collision Avoidance Works" (*Iris Automation*, 23 September 2020)

such as radars, ultrasounds and LiDAR, which emit a signal that is then reflected by an obstacle and detected again by the sensor, or passive sensors, such as visual and infrared cameras that detect a signal given off by the object itself.[86] It is expected that such technologies will pave the way for the frequent use of UASs in BVLOS mode and will eventually lead to their fully autonomous operation without any human intervention during flight.[87] This full level of autonomy is not currently permitted as the UK CAA requires that during BVLOS operations in non-segregated airspace, the remote pilot shall always be "capable of immediately taking active control of the UA".[88]

With respect to BVLOS operations in segregated airspace, the UK CAA has the authority to establish permanent or temporary Dangerous Areas (DAs).[89] The UAS would have the exclusive use of the DA, yet the CAA requires that communications between the remote pilot and the Air Traffic Control providers are maintained to safeguard the safety of aircraft that operate in the vicinity of the DA or inadvertently enter it.[90]

It is noteworthy that the retained EU legislation and the ANO do not apply to indoor operations, namely "flights within buildings, or within areas where there is no possibility for the unmanned aircraft to 'escape' into the open air", such as tank ballasts, cargo holds and so on.[91] The ANO covers mainly aircraft "in flight" in the open air, which in case of a UAS is the period from the moment when it first moves for the purpose of taking off, until the moment when it next comes to rest after landing.[92] For cases of civil UAS operations not covered by the aviation legislation, the Health and Safety Executive may be competent, and a Memorandum of Understanding has been signed between the CAA and the HSE on division of competences and collaboration.[93]

While the regulatory framework requires constant evolution to reflect technological advancements in the field of UASs, the recent European regulations (that are retained in the UK) are flexible enough to accommodate the technological growth of commercial UASs. With BVLOS operations currently constituting the Holy Grail for several commercial UAS users, the Regulations leave significant leeway to the UK CAA whether to authorise them. However, whether permission is forthcoming will depend very much on the technology proving that it can reliably replace pilots in avoiding collisions and provide

---

www.irisonboard.com/detect-and-avoid-how-airborne-collision-avoidance-works/ (accessed 18 February 2021). See also J. Peskett, "Making Flying Safe: The Evolution of DAA in Unmanned Drones" (*Commercial Drone Professional*, 13 December 2020) www.commercialdroneprofessional.com/in-depth-automation-in-unmanned-drones-and-making-flying-safe/ (accessed 18 February 2021).

86 Ibid.

87 UK CAA, "Beyond Visual Line of Sight in Non-Segregated Airspace. Fundamental Principles & Terminology" (October 2020) https://publicapps.caa.co.uk/docs/33/CAP%201861%20-%20BVLOS%20Fundamentals%20v2.pdf (accessed 18 February 2021).

88 CAP 722, op. cit., no. 3, [2.4.2].

89 See UK CAA, "Policy for Permanently Established Danger Areas and Temporary Danger Areas" (21 July 2020) http://publicapps.caa.co.uk/docs/33/Policy%20Statement%20Permanently%20Established%20Danger%20Areas%20and%20Temporary%20Danger%20Areas.pdf (accessed 18 February 2021).

90 CAP722, op. cit., no. 3, [2.4.3].

91 Ibid., [1.2.3.4.1].

92 ANO, op. cit., no. 79, art. 3(b).

93 Memorandum of Understanding between the Health and Safety Executive, Health and Safety Executive Northern Ireland and the Civil Aviation Authority for aviation industry enforcement activities (December 2016) www.caa.co.uk/uploadedFiles/CAA/Content/Standard_Content/Our_work/About_us/Files/HSE%20CAA%20Memorandum%20of%20Understanding.pdf (accessed 18 February 2021). It is the intention of these authors to explore health and safety issues arising from the use of UAS in enclosed spaces in a subsequent publication.

for safe autonomous operations. With this in mind, we now move to examine the accident rates of commercial UASs in the UK and to what extent the current liability and insurance paradigms, that are not designed with UASs in mind, require reconsideration.

6.4 Insurance considerations for commercial unmanned aircraft systems

6.4.1 Accident rates of commercial unmanned aircraft systems in the UK

The official statistics demonstrate that the accident rates of commercial UASs in the UK are becoming a concern. Between February 2015 and July 2019, 59 commercial UAS accidents, namely incidents that led to damage to the UAV due to collision with water or ground, were reported to the UK Air Accident Investigation Branch (AAIB), 52 of which involved UASs of less than 20 kg.[94] Of these 59 accidents, 37 were the result of technical faults in the UAS, while only eight were caused by human error.[95] The remaining accidents were a combination of Controlled Flights Into Terrain (five) resulting from the impact of wind and/or rain, mid-air collisions (three) and other causes which involved the incorrect fitting of propellers (six).[96] From August 2019 to September 2020, 13 accidents were reported to the AAIB, with the vast majority (nine) caused by technical faults and only one caused by human error.[97] In total, 72 commercial UAS accidents have been reported in the last five-and-a-half years, averaging slightly over 13 accidents per year.

The AAIB has investigated one accident caused by a UAS used in maritime operations. The UAV in question was used to inspect a vessel in the North Sea and lost signal at approximately 30 metres from the vessel, crashing into the sea. The UAV was retrieved, yet the causes of the loss remain undetermined as the tests were inconclusive.[98] No personal injuries or property damage, other than the damage to the UAV, were reported.

These statistics highlight that malfunction of technology rather than human factors is the main source of commercial UAS accidents. This is not surprising considering that new transport technologies usually achieve operational maturity via a trial-and-error process. What distinguishes the development of UAS technology is that this process has not resulted in significant number of fatalities. While this might be the result of commercial UASs operating in controlled environments, it is a clear indication that UAS technology has, so far, effectively managed its side-effects. It is also a significant reason for the trust that commercial users and insurers have already placed in UAS operations.[99]

It is inevitable that the estimated exponential growth of commercial UASs will eventually lead to personal injuries and/or more significant property damage. Admittedly, UASs do not have the catastrophic potential of aircraft; yet their proximity to the ground poses significant risks of third-party damage. Their relatively small size is deceptive as the AAIB,

94 For the purposes of our analysis in this Chapter we reviewed the official accident reports of the UK AAIB for UAS, and the 2018 and 2019 Annual Safety Reports of the UK AAIB and the UK AAIB Bulletin 1/2020 that are available at www.gov.uk/aaib-reports? (accessed 18 February 2021).

95 Ibid., AAIB Bulletin 1/2020, Table 3.

96 Ibid.

97 Ibid., AAIB accident reports' database available at www.gov.uk/aaib-reports? (accessed 18 February 2021).

98 See AAIB investigation to Falcon 8 Trinity Asctec (17 April 2017) available at www.gov.uk/aaib-reports/aaib-investigation-to-falcon-8-trinity-asctec-uas-bas-02 (accessed 18 February 2021).

99 See, op. cit., Chapter VII.

using the DROPS calculator, estimated that a "blunt object with a mass of 4.97 kg falling from a height of more than about 3 m could result in a fatal injury to someone wearing a hard hat".[100]

As such, the financial consequences of third-party liability resulting from the operation of UAS requires its reallocation via insurance. In the next sections, we examine the application of Regulation 785/2004 (as amended) on mandatory insurance for aircraft operators on UASs, and we also review a standard UAS insurance policy that was recently made publicly available.

6.4.2 Minimum insurance requirements for operating unmanned aircraft systems

Regulation (EC) 785/2004 on insurance requirements for air carriers and aircraft operators (as amended) imposes minimum insurance requirements to operators of aircraft in the EU.[101] The UK has retained this Regulation with necessary amendments, imposing the same insurance requirements to aircraft operators in the UK.[102] Regulation 785/2004 does not directly refer to UASs; yet its provisions are flexible enough to accommodate them. It imposes a requirement on aircraft operators to purchase insurance to cover their liabilities arising from the use of aircraft. While the Regulation does not provide a definition of aircraft, commercial UASs are covered by it to the extent that they do not fall into one of the Regulation's exclusions, which is unlikely in the case of commercial UASs.[103]

Regulation 785/2004 requires insurance to be purchased for each and every flight of a UAV,[104] which can be done by means of a "per flight/daily/weekly/monthly or annual policy".[105] The insurance is to be purchased by the aircraft operator which is defined as "the person or entity . . . who has continual effective disposal of the use or operation of the aircraft".[106] The presumption is that the registered owner is the operator, unless it can prove that someone else was operating the UAS in which case the obligation moves to the actual user.[107] This delegation would occur when the registered owner of the UAS leases the UAS to an operator without any remote pilots or visual observers (dry lease). In such instance, the insurance obligation falls on the person or entity that operates the UAV. In cases where the owner leases the UAS together with the remote pilot and observers (wet lease), the

100 UK AAIB Bulletin 1/2020, at 25 available at www.gov.uk/aaib-reports? (accessed 18 February 2021).

101 Regulation (EC) No. 785/2004 of the European Parliament and of the Council of 21 April 2004 on insurance requirements for air carriers and aircraft operators OJ L 138, 30.4.2004, at 1–6 as amended by Commission Regulation (EU) No. 285/2010 of 6 April 2010 amending Regulation (EC) No. 785/2004 of the European Parliament and of the Council on insurance requirements for air carriers and aircraft operators OJ L 87, 7.4.2010. at 19–20 and further amended by Commission Delegated Regulation (EU) 2020/1118 of 27 April 2020 amending Regulation (EC) No. 785/2004 of the European Parliament and of the Council on insurance requirements for air carriers and aircraft operators OJ L 243, 29.7.2020, at 1–2.

102 Retained Regulation No. 785/2004 on insurance requirements for air carriers and aircraft operators.

103 For the exemptions from its scope, see Regulation 785/2004, op. cit., no. 101, art. 2(2).

104 Regulation 785/2004, op. cit., no. 101, art. 4(2).

105 UK CAA, "Permissions and exemptions for commercial work involving small unmanned aircraft and drones" at www.caa.co.uk/Commercial-industry/Aircraft/Unmanned-aircraft/Small-drones/Permissions-and-exemptions-for-commercial-work-involving-small-unmanned-aircraft-and-drones/ (accessed 18 February 2021).

106 Regulation 785/2004, op. cit., no. 101, art. 3(c).

107 Ibid.

obligation to insure remains with the owner. This would be the case even if the owner receives instructions from the lessee regarding the scope of the performed work.

The obligation to insure covers any UAV flight "within, into, out of, or over the territory" of an EU Member State,[108] or of the UK under the UK Regulation.[109] As such, Regulation 785/2004 covers domestic UAS flights, irrespective of their duration or distance flown. It also provides that aircraft operators are required to purchase insurance to cover their "aviation-specific liability in respect of passengers, baggage, cargo and third parties".[110] With no commercial availability in terms of flying passengers and baggage yet, UAS operators are required to maintain third-party liability insurance. Such cover is required to include liability for "death, personal injury and damage to property caused by accidents"[111] and also war and warlike risks, such as "acts of war, terrorism, hijacking, acts of sabotage, unlawful seizure of aircraft and civil commotion".[112]

The requirement to purchase insurance to cover liability in respect of loss, damage, destruction or delay of cargo is controversial regarding the use of UASs to transport goods. The Regulation imposes a requirement to have cargo insurance of at least 22SDRs per kg,[113] reflecting the liability limits of air carriers under the Montreal Convention 1999.[114] The application of this requirement on deliveries by means of a UAS is questionable as the Regulation does not define the term "cargo" with the prevailing judicial interpretation under the Montreal Convention 1999 requiring the issuance of an air waybill.[115] As UAS deliveries to vessels are currently undertaken without the issuance of any documents of transport, it is arguable that operators are not subjected to the cargo-related provisions of the Regulation.

This issue also raises a broader question, namely whether it is justifiable to impose mandatory cargo insurance requirements for the carriage of goods by commercial UAVs. The main reason behind the cargo insurance requirement of the Regulation is to protect the consumer for the loss or damage of his/her goods during transportation. In a different work we have argued that most air cargo is currently moved by either institutional consignors or freight forwarders which do not require such protection, as they commonly purchase first-party insurance.[116] Reflecting such practices, major trading countries, such as the USA, do not impose a cargo insurance requirement. It is suggested that this philosophy is more appropriate in the context of commercial UASs as it allows parties to make alternative insurance arrangements rather than streamline the risk of transport via the UAS operator. Inevitably, when UAS consumer deliveries become a reality, mandatory cargo insurance might be a viable option. Nevertheless, it is suggested that the current one-size-fits-all approach requires reconsideration.

---

108 Regulation 785/2004, op. cit., no. 101, art. 2(1).

109 Retained Regulation 785/2004, op. cit., no. 102, art. 2(1).

110 Regulation 785/2004, op. cit., no. 101, art. 4(1).

111 Ibid., recital 14.

112 Ibid., art. 4(1).

113 Ibid., art. 4(1).

114 Ibid., recital (6); Convention for the Unification of Certain Rules Relating to International Carriage by Air, agreed at Montreal on 28 May 1999.

115 D. Mcclean et al., *Shawcross & Beaumont: Air Law* (Lexis Nexis, 2020), Division VII, Chapter 34, [901].

116 G. Leloudas, "The Future Challenges of Airlines' Liability Insurance Law" in Brian Havel and Jeremias Prassl (eds), *Elgar Research Handbook on International Aviation Law* (Elgar Publishing, forthcoming).

With respect to third-party liability, the amount of minimum insurance required reflects the MTOM of the UAV.[117] In the case of UAVs used in maritime operations it is expected that most of them fall into the first category of cover, namely 750,000 SDR per flight, as they have MTOM of less than 500 kg.[118] Even when the weight of the payload is added, it is unlikely that their weight will exceed the threshold of 500 kg any time soon.

The Regulation requires that third-party liability cover is provided on a per occurrence (that is, per accident) basis.[119] Yet it permits cover against third-party war risks to be provided on an aggregate basis.[120] The reason behind this exception is that the aviation insurance market does not have the capacity to cover such risks on per-accident basis following the events of 9/11.[121] It is questionable whether this indulgence should be extended to UASs. On the one hand, the practice of the airline insurance market is linked to the catastrophic potential of commercial aircraft as witnessed by the events of the 9/11 attacks. To avoid the recurrence of the insurance crisis that followed 9/11, third party war risk cover to commercial airlines is provided by means of several layers of excess insurance and is only available on an aggregate basis. On the other hand, UASs, albeit small, can be used to carry weapons of mass destruction and other materials that have the potential to cause catastrophic damage especially if used against vulnerable targets, such as nuclear energy plants. Considering that security arrangements around the use of UAS in prohibited zones are under development, the insurance market takes a precautionary approach and provides cover against third-party war risks of UAS operations on an aggregate basis.

Overall, we believe that the Regulation does not require major amendment to accommodate the use of commercial UASs. Certainly, some adjustments are required as there is little doubt that specific reference to UASs and an amendment of the definition of the term "flight" to accommodate rotary-wing UASs would be an improvement as they would leave no doubt that the Regulation applies to them.[122] Yet, there is no urgent need to overhaul it.

6.4.3 Insurance policies for unmanned aircraft systems

Regulation 785/2004 imposes minimum insurance requirements for licensing purposes. In the case of commercial airlines, the amount of cover purchased is a multiplication of the minimum requirement reflecting their liabilities to passengers and third parties. In a similar manner, the UAS minimum cover of 750,000 SDR for third-party risks does not reflect the damage potential of UASs involved in maritime and offshore activities. As a result, the offshore industry has developed a practice of requiring $US 10 million third party liability insurance from UAS operators.[123]

---

117 Regulation 785/2004, op. cit., no. 101, art. 7(1).

118 Ibid., Category 1 of the Table in art. 7(1).

119 Regulation 785/2004, op. cit., no. 101, art. 7(1).

120 Ibid.

121 Ibid.

122 Art. 3(c) of Regulation 785/2004, *supra* no. 101 defines "flight" for third party liability purposes as "the use of an aircraft from the moment when power is applied to its engines for the purpose of taxiing or actual take-off until the moment when it is on the surface and its engines have come to a complete stop".

123 OGUK, *Unmanned Aircraft Systems Operations for Offshore Installations* – Guidelines (December 2019), [2.1.1.] https://oilandgasuk.co.uk/product/unmanned-aircraft-systems-uas-operations-management-standards-and-guidelines/ (accessed 18 February 2021).

The insurance market has embraced the use of commercial UASs, and several insurers currently offer UAS-related cover in the UK on an annual, monthly or pay-as-you fly basis with insurance limits available of up to £ 50 million per event. Insurers have the capacity to provide cover for BVLOS operations, delivery operations and other niche commercial activities.[124] Demand for UAS insurance is such that it is expected that the London insurance market will eventually produce a standard policy for UASs. For the purposes of our analysis in this chapter, we will make use of Flock's Fly Unlimited insurance policy for commercial UAS operators.[125] The aim of this section is not to provide a comprehensive analysis of the published policy, but to identify certain issues of concern regarding the insurance cover for commercial UAS operations. The Fly Unlimited UAS policy is based on AVN1D, which is the standard insurance policy covering the hull and liability risks of general aviation operators.[126]

As expected, the Fly Unlimited policy does not have a passengers' liability section and is divided into three main sections. Section I covers loss or damage of the UAS, section II covers third party liability caused by the operation of the UAS, while section III contains the exclusions, conditions precedent, general conditions, definitions and extensions of cover.

Section I provides cover against the risk of physical loss or damage to the UAS and its payload whilst in flight, on the ground or in transit, up to the insured value.[127] Flight is defined as from "the time the UAV is switched on, moves forward in taking off or attempting to take off, whilst in the air, and until the UAV completes its landing run".[128] This definition resembles the definition of flight in AVN1D that applies "from the moment pressure is applied for take off until the landing run ends".[129] To cater for rotary-wing UAVs an alternative definition is provided, whereby the UAV is in flight from the moment "the UAV is switched on and the rotors are in motion as a result of engine power, the momentum generated therefrom, or autorotation".[130] The "rotary" wing definition requires the UAV to be switched on *and* the rotors to be in motion for the flight cover to apply, while the general definition suggests that flight cover starts when the UAV is switched on even though there is no power applied for moving forward. The former definition seems to create a clearer dividing line between "in flight" and "ground" cover, which attract different premium rates.

Unlike AVN1D, the Fly Unlimited policy covers, by means of an endorsement, physical loss or damage to the UAS while it is in transit.[131] No definition of "in transit" is provided, but the policy excludes cover under this endorsement when the UAV is in flight, or on the runway after cleared for take-off or before it reaches the runway exit after landing.[132] It also provides cover for physical loss or damage to the UAS while it is stored in an unattended

124 See, for example, the several types of cover for commercial UAS operations provided by Flock/Allianz at https://flockcover.com/ (accessed 18 February 2021).

125 In file with the authors.

126 AVN 1D – Aircraft Insurance Policy (30 August 2016) https://iua.co.uk/IUA_Member/Clauses/IUA_Member/Clauses/eLibrary/Clauses.aspx?hkey=6f7dd1a3-6ab3-4b10-94c2-5a8c644b1c32 (accessed 18 February 2021).

127 Fly Unlimited policy, op. cit., no. 125, section I, cl. 1a.

128 Ibid., section III-E.

129 Definitions' section, cl. 5 – AVN1D.

130 Fly Unlimited policy, op. cit., no. 125, section III-E.

131 Ibid., Transit Extension Endorsement (Attachment 10).

132 Ibid., cl. 3.1 of the Transit Endorsement.

vehicle (and also covers damage to the unattended vehicle itself), provided the UAS is kept out of sight and all security measures on the vehicle are in force.[133] Having said that, the following two conditions are to be satisfied for cover under the transit endorsement to be provided. The first one is that the UAS is packed and unpacked for transit by professional packers and according to the manufacturer's guidelines.[134] The second one is that all fire alarms and security measures that are notified to the insurer are engaged whenever the UAS is left unattended during transit.[135]

The cover against the risk of theft of the UAS is far more restrictive than the equivalent cover under AVN1D. While AVN1D does not impose any special conditions if an aircraft is stolen by third parties, theft of a UAS is excluded, unless i) the theft is the result of forcible entry or exit by third parties from its place of storage, namely a building, a container, a gated compound or a motor vehicle; ii) the security measures are in force at the time of the theft; and iii) the UAS is kept out of sight at all times.[136] The wording is restrictive enough to exclude the clandestine theft of UAS from a warehouse of the operator. It is not surprising that insurers do limit cover for theft as UASs are susceptible to it, especially during transit, because of their size.

In terms of third-party liability the policy provides cover

> for all sums which the Insured shall become legally liable to pay, and shall pay, as compensatory damages (including costs awarded against the Insured) in respect of accidental bodily injury (fatal or otherwise) and accidental damage to property caused by the UAV described in the Schedule or by any object falling therefrom.[137]

The policy provides cover for "bodily injury" without further defining it. Airline policies adopt an expanded definition of "bodily injury" to cover mental injuries,[138] while AVN1D defines bodily injury in a restrictive manner requiring the bodily manifestation of the mental injury.[139] It might be argued that insurers leave determination of what constitutes "bodily injury" to the policy's applicable law, yet for the avoidance of doubt a definition would be advisable, considering that the term "bodily injury" is a controversial one in aviation practices.

Third-party cover is excluded in situations where the UAS causes "loss of or damage to any property belonging to or in the care, custody or control of the Insured".[140] It is arguable that a vessel undergoing an inspection by a classification society using UAVs might be considered under the control of the UAS operator. If this is the case, any liability for damage to the vessel during the inspection is excluded from cover, leaving hull insurers to provide cover for the loss.[141]

---

133 Ibid., cl. 4.3 and cl. 3.4 of the Transit Endorsement.

134 Ibid., cl. 4.2 of the Transit Endorsement.

135 Ibid., cl. 4.3 of the Transit Endorsement.

136 Fly Unlimited policy, op. cit., no. 125, section I, cl. 2.b.

137 Ibid., section II, cl. 1. The policy also covers damages caused by the invasion of privacy, but, as we mentioned, this issue is outside the scope of our analysis.

138 For example, an airline policy would define bodily injury to cover "physical injury, mental anguish, fright, shock, sickness, disease. Including death resulting therefrom".

139 Cl. 10 in the Definitions' section defined bodily injury as "bodily injury, sickness or disease including death at any time resulting therefrom".

140 Fly Unlimited policy, op. cit., no. 125, section II, cl. 2.

141 See, op. cit., Chapter VI.A.

As per standard aviation insurance practice, the Fly Unlimited policy contains an AVN52E endorsement that provides war risk liability cover for property damage and personal injury to third parties.[142] The cover is provided on an aggregate basis, as per airline practice, on the basis that the payload of a UAS has the potential to cause catastrophic losses.[143] It has to be noted that the endorsement does not provide war risk cover for hull damage to the UAS and/or its payload and separate cover is to be acquired.

Finally, the Fly Unlimited policy contains AVN124, the standard data event clause, that is becoming increasingly popular with airline placements as well. The clause essentially excludes from cover "loss, damage, expense or liability arising out of a Data Event", which is defined as "access to, inability to access, loss of, loss of use of, damage to, corruption of, alteration to or disclosure of Data".[144] The clause aims to exclude cover for the financial implications of stealing data from the insured or accessing its IT network for ransom. However, it clarifies that the policy covers physical damage to the UAS and/or bodily injury and/or property damage to third parties caused by the data breach.[145] That would be a situation where a cybercriminal tampers with the operating system of the UAS to get access to information leading to the crash of the UAS on property or people on the ground.

It is likely that the increased use of UAS will accelerate the frequency of accidents both in a marine and a broader commercial context. Yet it is unlikely that the increase will cause any significant disruption in the commercial UAS insurance market that seems to be confident in the reliability of the technologies that are currently used. Despite the absence of extensive historical data on UASs and their users, insurers are building this confidence by resorting to telematics that collect real-time information on the behaviour of UAS users and the reliability of their equipment, and the conditions under which UASs are operated, such as weather, location, route conditions and so on. Having access to such information enables them to adjust the premium (even on a per flight basis), offer rewards for safe operations, provide alternative routes for operating UAS based on the prevailing conditions and have a clear picture of the operating capabilities and deficiencies of several UASs.[146] In that respect, commercial UAS insurers have embraced the nascent technology by playing an active role in the operations of their insureds and in rating the related technologies from an early stage of development. This active involvement has been achieved despite (or in spite of) a liability system for third parties damage/personal injury that is not designed to cater for the needs of UAS operations and to which we turn our attention in the next section.

6.5 Liability considerations for unmanned aircraft systems

There is no dedicated liability system governing third-party liabilities arising from the operation of UASs. What has become common practice is to apply the rules dealing with

142 For a comprehensive analysis of war risk cover see K. Posner, T. Marland and P. Chrystal, *Margo on Aviation Insurance* (4th edn, Lexis Nexis 2014), Ch 19.

143 See, op. cit., Chapter IV.B.

144 Fly Unlimited policy, op. cit., no. 125, attachment 12.

145 Ibid., cls. 1 and 2.

146 S. McGee, "Drone Insurer Expands with Connected Fleet Product for Larger Companies" (*Insurance Times* 25 June 2019) www.insurancetimes.co.uk/news/exclusive-drone-insurer-expands-with-connected-fleet-product-for-larger-companies/1430751.article (accessed 18 February 2021).

property damage and personal injuries caused on the ground by the operation of aircraft. These rules are usually of domestic origin, as there has been little success in creating international rules on third party liability of aircraft operators.

The international regime regulating the liability of aircraft operators for third party damage on the ground consists of the 1952 Rome Convention (RC52)[147] as amended by the 1978 Montreal Protocol (MP78).[148] Yet it has received limited international acceptance, with only 51 States having ratified the RC52 and 12 States the MP78.[149] Furthermore, the two international conventions on third party liability that were drafted in the aftermath of 9/11 to modernise the RC52 regime, namely the 2009 Montreal General Risks Convention[150] and the 2009 Montreal Unlawful Interference Compensation Convention,[151] are yet to be ratified and it is unlikely that they will be any time soon.[152]

The reluctance of States to commit to an international solution regarding liability of aircraft operators for ground damage has been attributed to the low limits of liability of the RC52, which remained unrealistic even after their increase by the MP78.[153] With several States adopting a system of strict and unlimited liability, accepting an international system providing for limited liability would provide lesser protection to third parties who are usually innocent bystanders.

Furthermore, States tend to regard the issue of third-party liability of aircraft operators as a domestic one, even when aircraft cause damage during international flights. Their argument is that in ground accidents the preponderance of evidence is located within the jurisdiction and, as such, there is little need to apply an international framework harmonising what is essentially a domestic issue. The 2009 Montreal Conventions were drafted under the pressure of the catastrophic damage witnessed at the events of 9/11, yet States were unwilling to ratify them once the memory of the events faded. This is the case even though the 2009 Montreal Conventions provide far superior protection to victims on the ground than the RC52 and the MP78.

In the case of UASs, the urgency for a unified third-party liability system is less pressing. Currently, commercial UAS services are not offered on a cross-border basis, and any accidents are treated as domestic events. UASs do pose an increased risk of ground damage as they operate closer to people and ground structure; yet it is arguable that domestic laws can manage this risk and an international solution would not offer better protection.

147 The Convention on Damage Caused by Foreign Aircraft to Third Parties on the Surface, Signed at Rome, on 7 October 1952.

148 Protocol to Amend the Convention on Damage Caused by Foreign Aircraft to Third Parties on The Surface, Signed at Rome on 7 October 1952, Signed at Montreal, on 23 September 1978.

149 For a list of States see ICAO, "Current Lists of Parties to Multilateral Air Law Treaties" available at www.icao.int/secretariat/legal/Lists/Current%20lists%20of%20parties/AllItems.aspx (accessed 18 February 2021).

150 Convention on Compensation for Damage Caused by Aircraft to Third Parties, done at Montréal on 2 May 2009.

151 Convention on Compensation for Damage to Third Parties, Resulting from Acts of Unlawful Interference Involving Aircraft, done at Montréal on 2 May 2009.

152 The General Risks Convention requires 35 ratifications to come into force with 12 States having ratified it as of February 2021. The Unlawful interference convention requires 35 ratifications by States whose total number of passengers departing from their airport exceed 750,000,000 on an annual basis. As of February 2021 only 9 States have ratified the Convention.

153 MP78, op. cit., no. 148, art. 11: The operator of a B787–800 with a MTOM of approx. 227,000 kgs bears liability of up to 15,305,000 SDRs (approx. 22,106,000 USD) with a sub-limit of 125,000 SDRs (approx. 180,500USD) per person injured or deceased. The operator faces unlimited liability if the third party proves that the damage was caused by a deliberate act or omission of the operator, its servants or agents – art. 12(1).

In the UK, the liability for accidents resulting in death, personal injury or property damage caused by the operation of UAS is determined by the liability scheme of s. 76 of the Civil Aviation Act 1982. This is designed to deal with accidents on the ground involving aircraft and does not address UASs per se. Even in 1982, it was not an innovative provision as it essentially was copied (with minor amendments) from s. 9 of the Air Navigation Act of 1920.

Its historical endurance is explained by its simplicity. Section 76(2) provides for a system of strict and unlimited liability when "material loss or damage is caused to any person or property on land or water by, or by a person in, or an article, animal or person falling from, an aircraft while in flight, taking off or landing". It is a system of strict liability as the victim is not required to prove the negligence or intention of the owner of the UAS and is only required to prove that personal injury or property damage was caused by an object or person falling from the UAV.[154] The only available defence to the owner of the UAV is the contributory negligence of the victim.[155] The Act makes no provision for special defences, such as acts of terrorism, hijackings or cyber interference, and creates an almost absolute liability system as it unlikely for the victim's injury to have been caused as a result of the victim's interference with the operation of the UAS.[156]

Despite its age, the provision is drafted in such a wide manner that, in principle, it can accommodate incidents arising from the use of UASs in a maritime context. It applies to damaged structures on water, thus covering damage to vessels and oil rigs.[157] The wording also leaves no doubt that personal injuries to persons on board vessels and oil rigs are equally covered as it applies to "any" person, irrespective of their location. It also applies to damage or personal injuries not only caused by the UAV itself, but also by any person or article falling from it.[158]

While UAVs are not carrying passengers yet, the wording would cover situations where the injury or damage is caused by the falling payload of the UAV, whether it is a camera, a sensor for structural surveys or vessel supplies, or by any propeller detaching from the body of the UAV. However, it is unlikely that Act applies to damage or injury caused by the remote pilot station rather than the UAV itself as the ANO defines small, unmanned aircraft to include "articles or equipment *installed in or attached to the aircraft* at the commencement of its flight".[159]

Furthermore, s. 76(2) provides that the victim can recover for "material loss or damage", which is further defined in s. 105 to include "loss of life and personal injury". The term "personal injury" has a relatively clear understanding in that it includes psychiatric injuries (at least in 1982 when the Act was enacted) and is distinguished from the term bodily injury. The argument that the use of the adjective "material" in the Act does not permit recovery for stand-alone psychiatric injuries was rejected in the case of *Glen v Korean Airlines Co Ltd.*[160] The defendants argued that the term "material" reduced the scope of

154 In that respect, the victim is not required to prove that the owner of the UAS recklessly or negligently caused the UAS to endanger persons or property on the ground as per s. 241 ANO, op. cit., no. 79.

155 Civil Aviation Act 1982, s. 76(2).

156 One such unlikely scenario might arise if the UAS injures the person that attempts to "jam" its operation.

157 The provision does not apply to flight inside enclosed spaces, such as tanks.

158 Civil Aviation Act 1982, s. 76(2).

159 ANO, op. cit., no. 79, s. 94(4).

160 EWHC 643 (QB), [2003] QB 1386.

the term personal injury as in 1920, when this provision had been originally drafted, it had referred to "physical or bodily damage or loss to the exclusion of mental or physical damage or loss".[161] The Court agreed that this had been the prevailing interpretation in 1920; yet it rejected the historical interpretation of an Act of Parliament (distinguishing them from international conventions) as "Acts must be construed in the light of contemporary circumstances", embracing developments in scientific understanding, "social conditions, technology, medical knowledge and in the meaning of particular words".[162] As such, it enabled, in principle, the recovery of stand-alone psychiatric damages, although it refused recovery of the claimants in question as they did not satisfy the common law test of being either primary victims or secondary victims with sufficient proximity and ties of love and affection with the injured.[163]

Liability under s. 76(2) attaches to the registered owner of the UAV, which creates a discrepancy with the mandatory registration requirement that is imposed on the operator of the UAV,[164] namely the entity that manages the UAS and is responsible for its maintenance, operation and ensuring that the remote pilot has the permissions to fly it. In that respect, classification societies that operate their own UASs would be required to register, while in the case of outsourcing the operation of UASs the responsibility to register lies with the outsourced operator of the UAS. Currently, this legislative discrepancy does not seem to cause issues as commercial UASs are usually operated by their owners even in cases where they are leased out. Such leases are usually arranged on a "wet" basis, namely that the owner leases the UAS together with the pilot and potential observers. As such, liability under s.76(2) falls on the operator of the UAS.

Having said that, it is likely that operating practices will change (as they did in the aviation industry), with owners of UASs leasing them out on "dry" basis, namely by providing the UAV with the lessee operating the UAV by using its own pilots and observers. In this situation, the arrangement of s. 76(2) has the potential to cause unfair results as it allocates the risk of liability to the owner that does not have navigational control over the UAS. Section 76(4) partly mitigates this issue as it makes the lessee liable in cases where the UAS has been leased out for more than 14 days without the provision of any "pilot, commander, navigator or operative member of the crew of the aircraft". In principle, this is a reasonable arrangement as it allocates the risk of liability to the person that operates the UAS.

However, this still leaves two issues unresolved. First, it does not cater for situations where the owner of the UAS provides visual observers or technical assistants, but not the remote pilot, to the lessee. Can a visual observer qualify as a navigator, disrupting the allocation of liability risks to the lessee? It is likely, although by no means certain, that they might qualify as such, considering that observers advise the remote pilot about the position of the UAS and any potential collision hazards. Much will depend on whether the remote pilot retains overall responsibility for the navigation of the UAS, which seems likely to be the case. The provision of a technical assistant is unlikely to change the allocation of risk

161 Ibid., [16].

162 Ibid., [24].

163 Ibid., [33] to [39]. The claimants witnessed the crash of a B747 shortly after take-off from Stansted airport and claimed for stand-alone psychological injuries.

164 For the registration requirements see, op. cit., Chapter III.B.3.

achieved as she/he will not, under normal circumstances, have a saying over the navigation of the UAS.

Second, the owner of the UAS remains liable for any third-party damage caused by the UAV during the first 14 days of the dry lease. This provision seems to have been copied from the RC52 which imposes a similar requirement to aircraft owners.[165] With the registration of UAS operators being mandatory, this is an unreasonable provision as it makes the owner/lessor potentially liable for a period that has no operational control over the UAS. The RC52 mitigates the impact of this arrangement by making the owner/lessor and the operator/lessee of an aircraft jointly and severally liable during this period, essentially enhancing the victim's protection who has the unilateral right to bring claims against both or either of them. However, the 1982 Act does not follow the same path. Instead, it gives the owner/lessor a right of indemnity against the lessee for any third-party claims that was required to satisfy because of the operation of a dry-leased UAS. A literal reading of s. 76(3) seems to absolve the operator/lessee from direct liability for the first 14 days which gives lesser protection to victims on the ground than the RC52. Despite the wording, we believe that the reasonable interpretation of s. 76(3) and (4) is that both the owner and the lessee remain potentially liable for the first 14 days of the dry lease and the owner is given the right to bring an indemnity action against the lessee in cases that he has satisfied claims arising from the operation of the UAS by the lessee.

While the creation of an international liability system to regulate the operation of UASs is not an urgent matter (for the reasons analysed above), it is worthwhile adjusting the liability system of s.76 to reflect the increased use of commercial UASs in the UK. We believe that the current system of strict and unlimited liability should remain in place, as it provides adequate protection to third-party victims, while it is not disruptive of the UAS insurance market. This balance might change if UASs were used for terrorism purposes or cause damage to critical infrastructures on the ground, yet there is nothing now to suggest that the liability system requires overhauling. Still, an adjustment is necessary to modernise the sharing of liability risks between owners and operators of UASs and to identify the legal status of visual observers and technical assistants.

6.6 Hull and P&I insurance for operating unmanned aircraft systems

Having examined the regulatory, liability and insurance implications of using UASs in a maritime context, it is time to examine the impact of their operation on marine insurers, namely hull insurers and P&I Clubs.

6.6.1 Hull insurance

Clause 6.2.5 of the Institute Time Clauses – Hull 1995 (ITCH 95) provides cover for loss or damage to the hull caused by contact with aircraft, helicopters or similar objects, or objects falling therefrom. The wording is broad enough to include damage caused to the hull by the operation of a UAS either when the UAS is used as part of the vessels' operations or

165 RC52, op. cit., no. 147, art. 3.

independently of it. The clause covers damage to the hull caused by the crash of the UAV on it or by any object, such as its payload or the propellers, falling on the hull structure.

The cover in cl. 6.2.5 is subject to the due diligence proviso of the ITCH 95, namely that the loss or damage "has not resulted from want of due diligence by the Assured, Owners, Managers or Superintendents or any of their onshore management". In broad terms, the due diligence proviso has the aim of preventing foul play from senior management, with the negligence of line employees such as the crew not subjected to it. For years, there was a debate under English law whether the proviso requires a recklessness or a negligence standard. The judgments in *Sealion Shipping Ltd v Valiant Insurance Co (The Toisa Pisces)*[166] and *Versloot Dredging v HDI-Gerling Industrie Versicherung AG (The DC Merwestone)*[167] resolved this debate by confirming that the proviso required merely negligence of the company's management. This conclusion ran against the then prevailing view in the insurance industry, which considered it to apply to "losses resulting from intentional acts . . . which are either reckless or show scant regard for proper standards of maritime safety".[168]

Significantly for the operation of UASs in a maritime context, both cases rejected an interpretation that would have required a proactive risk management approach, and meant that any failure of the crew or contractors created a *prima facie* case of failure of their technical supervisors in authorising or implementing safety processes and inspections. In that respect, insurers will be able to deny cover by using the proviso, only if they prove that the technical and safety managers were negligent in exercising their duties to "prepare or equip the vessel for the voyage or service she is about to perform".[169] As a result, it is unlikely that negligent or reckless acts and omissions of the UAS operator would be considered in breach of the "due diligence" proviso.

Still, a manager of the assured who supervised the operation of UAS might be in breach of the proviso in cases where the information received from the UAS operator was inadequate and clarifications would be required before authorising the operation. But even in that case, due diligence is not to be equated with the failure to take all necessary measures to prevent a loss, unless it was reasonable to take such a course of action on the basis of the information provided or the information it was reasonable to be provided at the time the decision was made.

For the proviso to take effect, insurers are required to prove that i) the shipowner's managers were negligent in setting or approving the policy in terms of safety and operation of the UAS; and/or ii) their response to the information they were getting from the operator of the UAS was sub-par considering their knowledge at that moment. The negligence of the operating crew of the UAS shall not create a *prima facie* case that the assured and its managers failed in their supervisory duties as this interpretation would require the reversal of the burden of proof.

---

166 [2012] EWHC 50 (Comm); [2012] Lloyd's Rep 252. The Court of Appeal confirmed the decision of the High Court in [2012] EWCA Civ 1625; [2013] 1 Lloyd's Rep 108 without discussing the "due diligence" proviso.

167 [2013] EWHC 1666 (Comm); [2013] Lloyd's Rep 131.

168 D. Sharp, *Upstream and Offshore Energy Insurance* (Witherbys Insurance, 2009), at 355.

169 J. Gilman QC et al., *Arnould: Law of Marine Insurance and Average* (19th edn, Sweet & Maxwell 2020), [23–72] argue, by reference to US cases, that this is the correct approach under English law.

The wording of the proviso in the ITCH 95 extends the list of relevant persons to superintendents or the assured's onshore management.[170] US cases suggest that even lower level onshore management would qualify.[171] Yet, we believe that this interpretation should not be followed in English law as, in most occasions, lower level shoreside employees do not have control over the company's safety decisions but exercise delegated duties regarding the operation of the vessel. Their inclusion would limit disproportionately the cover of the policy.[172]

Clause 2.1.9 of the International Hull Clauses – 1/11/2003 (IHC03) also provides cover for loss or damage caused to the hull by "contact with satellites, aircraft, helicopters or similar objects, or objects falling therefrom". While almost identical to the IHC95, the wording is not subject to the "due diligence" proviso which essentially makes recovery for damage caused by a UAS more likely.

Furthermore, both the ITCH 95 and the IHC 03 provide cover for loss or damage to the hull caused by the negligence of "Master Officers Crew or Pilots",[173] subject to the due diligence proviso analysed above. In case the UAS is operated by a member of the crew of the assured (rather than a contractor) this clause also has the potential to cover damage to the hull caused by a UAV. The operation of UASs has the potential to renew interest in the application of this clause as damage to the hull caused by UAS is linked to cases involving machinery damage where "negligent handling causes damage but no marine peril operates".[174]

One potential issue is whether hull clauses have the potential to cover damage to or loss of the UAS itself. With a UAS unlikely to qualify as part of the hull, the question is whether there are any extensions that might provide cover in such instances. In IHC 03 the leased equipment extension (cl. 3) provides as follows: "This insurance covers loss of or damage to equipment and apparatus not owned by the Assured but installed for use on the vessel and for which the Assured has assumed contractual liability, where such loss or damage is caused by a peril insured under this insurance."

It is unlikely that this clause provides cover for damage to UASs as they are not "installed on the vessel". Although the ITCH 95 do not contain such extension, it is common practice for a "leasing equipment" clause to be added along the lines of the IHC 03 clause. The result is the same: namely, that the hull policy does not provide cover for property damage to the UAS.

6.6.2 P&I insurance

At the time of writing the International Group of P&I Clubs confirmed that liabilities arising from the operation of UASs are covered by the Pooling Agreement.[175] However,

170 The original wording of the proviso covers the assured, the owners and the managers of the property insured.

171 J. Hill, *O'May on Marine Insurance: Law and Policy* (Sweet & Maxwell 2003), at 136–137.

172 See G. Leloudas and B. Soyer, "Standard Contracts Used in the Offshore Insurance Sector" in B. Soyer and A. Tettenborn (eds), *Offshore Contracts and Liabilities* (Informa Law from Routledge 2015), at 209, 238–240.

173 Cl. 6.2.2 of ITCH 95 and cl. 2.2.3 of IHC 03.

174 *Arnould*, op. cit., no. 169, [23–69].

175 B. Burkard and J. Hines, "The Use of Drones in Shipping and Cover Implications" (Standard P&I Club, Technology Bulletin, September 2018) https://standard-club.com/media/2767845/the-use-of-drones-in-shipping-and-cover-implications.pdf (accessed 18 February 2021).

the Group has reserved the right to reconsider its decision based on the accident rates of UASs.

By confirming cover, the Group treats UASs differently to mini submarines and diving bells, whose liabilities are excluded from cover.[176] For cover to be provided, the main prerequisite is that the UAS be used in direct connection with the operation of the ship.[177] For example, while UASs used in operations related to the port terminal will not fall into P&I cover,[178] liability incurred as a result of UAS operation in assisting or supervising the loading or unloading of the vessel at a port is likely to be covered.

Apart from liability arising from damage to cargo, the operation of UASs can trigger P&I cover for death or injury of crew and passengers on board the vessel. In the case of crew, cover is provided whether the UAS is operated by the injured crew, a different crew member or a contractor. With UASs increasingly being tested as a means of providing entertainment to passengers on cruise ships, it is likely that passenger claims for personal injury caused by UASs will increase. Having said that, recovery for passengers might not be straightforward as UAS-related accidents are unlikely to qualify as "shipping incidents" under the Athens Convention 1974 as amended by the 2002 Athens Protocol (Athens Convention).[179] For non-shipping incidents, passengers are required to prove that their personal injuries are caused by the negligence of the UAS operator.[180] As such, the Athens Convention creates a liability regime that is less favourable than the strict regime of s. 76(2) of the Civil Aviation Act 1982 and it is likely that the exclusivity principle of the Athens Convention will prevent passengers from bringing claims under the 1982 Act.[181] Having said that, fines imposed by the UK CAA for breach of UAS-related regulations and costs of formal inquiries undertaken in the aftermath of UAS accidents are not automatically recoverable under standard P&I cover; their recovery is left at the discretion of the Club.[182] In the case of fines the discretion is to exercised only if the member has satisfied the Club that it took reasonable steps to avoid the accident giving rise to a fine.[183]

P&I Clubs recommend that service contracts with UAS operators should contain a knock-for-knock clause or at least that "the member does not assume responsibility for liabilities that they would not otherwise have had at law".[184] Under a knock-for-knock clause,

---

176 S. Hazelwood and D. Semark, *P. &. I Clubs Law and Practice* (4th edn Informa Law), at [10–50].

177 See, e.g., Standard Club, *P&I Rules* (2020/2021), Rule 2.1 at www.standard-club.com/media/3256558/pi-rules.pdf (accessed 18 February 2021) and Gard, *Rules* (2021), Rule 2.4 at www.gard.no/Content/31177106/GardRules_2021_Web.pdf (accessed 18 February 2021).

178 Gard, *Guidance to the Rules 2021* (2021), Rule 2.4.a www.gard.no/web/publications/document/chapter?p_subdoc_id=20747953&p_document_id=20747880 (accessed 18 February 2021).

179 Athens Convention Relating to the Carriage of Passengers and Their Luggage by Sea, 1974, Concluded at Athens on 13 December 1974 and Protocol of 2002 to the Athens Convention Relating to the Carriage of Passengers and their Luggage by Sea, 1974, Adopted London 1 November 2002. A "shipping incident" under the Athens Convention is defined as a "shipwreck, capsizing, collision or stranding of the ship, explosion or fire in the ship, or defect in the ship", art. 3(5)(a).

180 Athens Convention, op. cit., no. 179, art. 3(2); see also B. Soyer and G. Leloudas, "Carriage of Passengers by Sea: A Critical Analysis of the International Regime" Michigan State International Law Review, 26 (2018): 483.

181 See Athens Convention, op. cit., no. 179, art. 14 which provides that "[n]o action for damages for the death of or personal injury to a passenger, or for the loss of or damage to luggage, shall be brought against a carrier or performing carrier otherwise than in accordance with this Convention".

182 See, e.g., Standard Rules, op. cit., no. 177, Rules 3.16 and 3.17 and Gard Rules, op. cit., no. 177, Rules 47 and 45 respectively.

183 Standard Rules, op. cit., no. 177, Rule 3.16 and Gard Rules, op. cit., no. 177, Rule 47.

184 B. Burkard and J. Hines, op. cit., no. 175.

the operator of the UAS agrees to cover claims for damage to its own property (essentially the UAS) or property of its contractors, as well as personal injury to its employees and/or contractors, even if such damage and/or injury is the result of the shipowner's negligence or breach of statutory duty. Similarly, the shipowner agrees to assume responsibility for claims for damage to its own property (the vessel) or that of its contractors, personal injury to its employees and/or contractors, even in situations where such damage and/or injury is the result of the UAS operator's negligence or breach of statutory duty.[185] Furthermore, the UAS operator agrees to indemnify the shipowner if he paid claims that should have been entertained by the UAS operator. Usually, a party cannot rely on the protection of the knock-for-knock agreement if the damage or injury is caused by its deliberate act, wilful misconduct or recklessness.[186]

In aviation the standard practice is to channel liabilities via the aircraft operator on the basis that it is at the best position to obtain insurance to cover the liabilities of its service providers. In a similar manner, standard UAS service agreements channel liability risks via the UAS operator.[187] They usually provide that damage to the UAS, as well as damage to property belonging to the service user and personal injuries to employees of the user are to be covered by the UAS operator, unless they are caused by the user's wilful misconduct or, under a less strict version, by its negligence.[188] At the same time, UAS services agreements often provide that damage caused to property belonging to a third party by the UAS "during the Flight or as a result of the performance of the Services" is to be covered by the operator, unless it is caused by the user's wilful misconduct or negligence.[189] With the threshold at wilful misconduct, the channelling of liability risks via the operator is near absolute, considering that wilful misconduct is a rather high threshold to satisfy.

Having said that, it is questionable whether the UAS operator is in better position than maritime users to get property and liability insurance for damage caused in the provision of services. With both the UAS sector and the UAS insurance sector in a developing state, it is often the case that the customer is in far better position to cover liabilities incurred by the operation of UAS. That can be achieved by obtaining UAS endorsements in their existing buildings and public liability insurance policies, with the cost of the additional premium having significantly lesser impact on them. At the same time, one of the arguments for channelling liability risks via an aircraft operator is that it enables service providers to offer lower prices. In the case of UAS operations such an argument would require the channelling of the risk via the maritime user, considering that the service is provided by the UAS operator rather than the other way around.

We believe that the requirement of P&I Clubs for knock-for-knock arrangements in the operation of UASs is more sensible considering the financial disparity between the two contracting parties and the damage potential of UAS. Especially in the offshore sector, it has been long argued that traditional indemnities, such as the ones described above, are of little use:

185 See F. Lerede, "Knock-for-Knock: The P&I Perspective" in B. Soyer and A. Tettenborn (eds), *Offshore Contracts and Liabilities* (Informa Law from Routledge 2015), at 201, 204 and S. Rainey, "The Construction of Mutual Indemnities and Knock-For-Knock-Clauses" in B. Soyer and A. Tettenborn (eds), *Offshore Contracts and Liabilities* (Informa Law from Routledge 2015), at 68.

186 Ibid.

187 In our analysis we have used a standard "Drone Services Agreement" provided by Lexis Nexis.

188 Ibid., cls. 5.1, 5.3, 5.4.

189 Ibid., cl. 5.2.

> if standard indemnity type risks were adopted in this sector, then it is unlikely that any contractor would have sufficient balance sheet, appetite or insurance coverage to take on the risk. The risks posed to the supply chain would probably be simply unacceptable and/or uninsurable and so the contractor could not back its risks off. If these risks were accepted, then the supply chain would be looking to pass the inevitable high insurance premiums to the owner, in turn driving up the cost of the works. In the "knock for knock" arrangement, because a party is only required to indemnify (and hence, insure against) the personal injury/death and damage to property and pollution risk in respect of its own employees and property, this makes the arrangement more practical for the types of risk experienced in oil and gas and offshore projects. This avoids exposure for the main contractor and the supply chain to the potentially enormous costs of claims in the worst-case scenario.[190]

For UAS operators, knock-for-knock clauses provide a superbly beneficial arrangement as their liability is limited to first-party losses, namely physical damage to the UAS and personal injuries to its pilots, observers and technical assistants. This arrangement enables the development of insurance products for UAS operators involved in high-risk activities and makes the maritime sector more appealing to UAS operators as the main liability exposures remain with the maritime user that is in superior position to manage them. At the same time, knock-for-knock arrangements have a broader effect. They essentially subsidise the development of a new, promising technology and replace the need for imposing a statutory limitation of liability that is unlikely to be agreed any time soon. Having said that, knock-for-knock arrangements do not negate the obligation of the UAS operator to obtain third party insurance as per the requirements of Regulation 785/2004.[191]

The decision of the International Groups of P&I Clubs to provide cover for liabilities arising from the use of UAS is a significant vote of confidence regarding their further deployment in the maritime industry. Taken together with the increasing availability of UAS-specific insurance,[192] it has wider repercussions as marine and general insurers send strong signals that the commercial UAS industry has entered an era of rapid expansion of the risks that it is willing to underwrite. In this expansion, it is hoped that knock-for-knock arrangements will become more common outside the constraints of the maritime industry as they give incentives for the UAS industry to develop. Yet, a more important factor to the further deployment of UAS is the building of trust with their users and the public to which we turn our attention below.

6.7 Risk perceptions of unmanned aircraft systems

The UAS industry is currently going through a trust-building exercise with the public and commercial users of UAS. The results are tentative, yet some patterns emerge. Business users of UAS in the UK seem to embrace their potential. In a recent PWC report 40% of 252 senior business decision makers responded that their business was expecting or planning to buy UAS services, 33% said that UASs were already being used effectively by their industry and 56% felt positive about them.[193]

190 S. Boggs, "Indemnities in Offshore Construction Projects – Do Not Be Shocked by Knock for Knock" (April 2016) www.squirepattonboggs.com/~/media/files/insights/publications/2016/09/construction-and-engineering-update-autumn-2016/construction-and-engineering-update-autumn-2016.pdf (accessed 18 February 2021).

191 See, op. cit., Chapter IV.B.

192 See op. cit., Chapter IV.C.

193 E. Whyte and J. Murray, "Building Trust in Drones" (PWC, April 2019) www.pwc.co.uk/intelligent-digital/drones/building-trust-in-drones-final.pdf (accessed 18 February 2021).

Regarding consumers, the results indicate either a neutral or a negative view of them. An extensive Australian survey demonstrated that more education of the Australian public was required about the use of UASs: "perceptions of the benefits and the overall acceptability of the technology are . . . neutral. The neutrality of the responses suggests that the public has yet to form an opinion in relation to UAS. The most likely contributor to this situation is the lack of knowledge held by respondents."[194] At the same time, the PWC report indicates that the UK public has a rather negative image of UASs, especially towards air taxis, package deliveries and air ambulances. Yet the use of UASs in search and rescue operations, police operation and in emergency situations are overwhelmingly supported.[195]

One might argue that consumer attitudes towards UASs are irrelevant in the context of maritime uses. Yet, it has rightly stated that commercial operators of UASs "should not shy away from public concerns, and allow an information vacuum to be filled with negative perceptions".[196] A major accident caused by a UAS in a maritime context that attracts media attention would have the potential to amplify negative social perceptions of the industry that could be addressed by regulators taking a stricter approach to licensing and the further development of BVLOS capabilities. The industry's state of development has not reached a point yet where commercial operations have acquired an independence from consumer-related operations in terms of regulation and any well-publicised accident has the potential to affect the industry in its entirety.

As the evolution of autonomous cars demonstrates, societal risk perceptions dictate that modern technologies develop in a proactive manner, factoring in the prevention of accidents from an early stage of development.[197] Otherwise, questions of acceptability of the new technology are raised, especially by the media that are quick to question the science behind their use by simplifying technical information and amplifying issues of liability.

The UAS industry is at a critical point of this trust-building exercise. It was already argued in 2014 that "today drones are at the top of the news. It's to the point where all this drone talk has become droning".[198] The preoccupation with them reflects that they are not a niche industry anymore used by a few enthusiasts but are at the verge of occupying a mainstream role at least regarding commercial operations. Their further evolution will depend on persuading the public that their benefits and the safety of operations justify their further integration into societal uses.[199]

6.8 Conclusion

The use of UASs in the maritime industry is on an ascending trajectory because of the operational and financial benefits they offer to several stakeholders. With the licensing environment in Europe and the UK having recently gone through an overhaul to accommodate

194 R. Clothier, D. Greer, D. Greer and A. Mehta, "Risk Perception and the Public Acceptance of Drones", Risk Analysis, 35 (2015) at 1167, 1178.

195 PWC Drones Report, op. cit., no. 193.

196 Ibid.

197 See M. Chatzipanagiotis and G. Leloudas, "Automated Vehicles and Third-Party Liability: A European Perspective" U Illinois Jo of Law, Technology & Policy (2020): 109.

198 E. Zintel, "Drone Talk is Droning" (*Editor & Publisher* 2014) as referenced in R. Luppicini and A. So, "A Technoethical Review of Commercial Drone Use in the Context of Governance, Ethics, and Privacy", Technology in Society, (2016) 46: 109, [5.10].

199 Ibid., Luppicini and So.

operational and technological developments, manufacturers of commercial UAS are now preoccupied with developing the technology that will facilitate the granting of licenses for BVLOS operations. In this process, the existing liability system for third party property damages and personal injuries in the UK does not create impediments to their further deployment. Although certain adjustments (identified in the paper) are required, there is no (urgent) need to change the liability paradigm. Similarly, there seems to be no urgent need to change the insurance paradigm, despite certain adjustments required, as both UAS and marine insurers have already embraced UASs and, via the use of technology, such as telematics, are able to provide financial certainty in the operation of UASs for commercial and maritime projects.

CHAPTER 7

The role AI and machine learning will play in maritime and trade law

Julian Clark and David Owens***

7.1 Introduction

This chapter is intended as a brief introduction to the future role of artificial intelligence (AI) and machine learning in maritime and trade law. It therefore seems appropriate to us to commence in Section 7.2 with a brief note on what we consider AI and machine learning to mean. Such technologies are more often used than defined, and not everything is as it may at first seem. Indeed much confusion is caused by application of the wrong base line definition and so we have taken this as our starting point.

In Section 7.3 we will then run through some of the interesting ways in which AI and machine learning is currently being developed in the maritime and trade worlds, before focussing on autonomous vehicles. In particular, we will look at how the legal concepts of seaworthiness and due diligence may be challenged by the introduction of autonomous vehicles. Then we will consider how autonomous vehicles may change the practice of casualty investigation – and what evidence gathering in such a situation will mean.

In Section 7.4 we will look at the role AI and machine learning will play in the practice of maritime and trade law, in terms of what it means to be a maritime and trade lawyer. We will consider some of the ways in which AI and machine learning will lead to the automation of certain elements of maritime and trade law; this will both reduce overall demand for lawyers, but also open new opportunities for lawyers to work in the development and application of the relevant technology, as well as creating the regulatory framework in which it will operate. We will then look at what the limits of such technology might be – can we be replaced by AI and machine learning altogether?

7.2 Introduction to AI and machine learning

A short note on the terminology used in this chapter may be helpful. In this chapter, we will discuss "The role AI and machine learning will play in maritime and trade law". This raises the question of what we mean both by machine learning, and AI. Although we do not believe there is a single, commonly accepted definition of either concept, a useful working definition might be the following from the Royal Society, the oldest national scientific institution in the world:

> Machine learning is the technology that allows systems to learn directly from examples, data, and experience.

* **Senior Partner, Ince**
** **Managing Associate, Ince**

DOI: 10.4324/9781003155195-7

> If the broad field of artificial intelligence (AI) is the science of making machines smart, then machine learning is a technology that allows computers to perform specific tasks intelligently, by learning from examples. These systems can therefore carry out complex processes by learning from data, rather than following pre-programmed rules.[1]

In other words, machine learning may be thought of as a narrower subset of AI in general. In machine learning, traditionally the system will be given a task, a set of data, and will "teach" itself through algorithms how to achieve the outcome. This may be done in various ways, for example:

i) In "supervised" machine learning, the system may be guided by humans in relation to certain inputs. So, for example, many readers will be familiar with computer-assisted disclosure processes, in which the system will be "trained" by observing a human reviewing a set of documents for relevance; the system will then attempt to acquire rules from those observations as to what documents may be relevant, and automatically apply those rules to the review of a larger set of documents. Image recognition software works in a similar manner. One supervised machine learning technique is the "random forest", in which the system creates a multitude of random potential decision trees (hence the "random forest"), each of which will be tried until one is found that "works". As discussed in Section 7.4, this technique is being used currently in certain systems seeking to predict the outcome of legal cases.
ii) In "unsupervised" machine learning, the system takes inputs that are not pre-assigned labels by human intervention. The system will then seek to create "clusters" of inputs that appear to the system to have common inputs to each other, but dissimilar to those from other clusters; or by "dimension reduction" to reduce the number of variables in a dataset by grouping similar or correlated attributes for better interpretation.
iii) In "reinforcement" machine learning, the system will be assigned a "reward" function, and seek through experience to optimise the path to that reward. This is the method used by AlphaGo Zero, the latest version of the program that has defeated the world's strongest professional Go players, which is said to have trained itself by playing 4.9 million games against itself,[2] thereby creating its own experience from which it can learn.

It will be noted from the loose description of AI ("making machines smart") that AI is an easier concept to recognise than define. It has been used to refer to a broad range of computational abilities, up to and including creating computers with human levels of cognition that can decipher commands, questions, statements and reply perhaps with wit and sarcasm (which clearly has not yet occurred). However, as a working definition of the difference in practical outcomes between machine learning and "true" AI, we would say that a machine

1 The Royal Society, *Machine Learning: The Power and Promise of Computers that Learn by Example*, (DES4702, 2017) para 1.1 at https://royalsociety.org/~/media/policy/projects/machine-learning/publications/machine-learning-report.pdf?la=en-GB&hash=B4BA640A1B3EFB81CE4F79D70B6BC234 (accessed 1 May 2021).

2 M. Kennedy, "Computer Learns to Play Go at Superhuman Levels Without Human Knowledge" (NPR, 18, October 2017) at www.npr.org/sections/thetwo-way/2017/10/18/558519095/computer-learns-to-play-go-at-superhuman-levels-without-human-knowledge (accessed 1 May 2021).

learning system takes knowledge, and uses that knowledge and learning from past interactions in order to make a prediction. "True" 2 AI, would take knowledge that it has, and could extrapolate it into a new situation.

So, for example, a machine learning system may notice that there is a strong correlation between rain in London and umbrella use in humans in the area. Having done so, when the system notices it is raining outside, it will suggest that the human takes an umbrella outside. A "true" AI system, on the other hand, may also notice the correlation between rain and umbrella usage. The system may then speculate that the umbrella is used as some form of protective mechanism, and suggest that the human takes the umbrella outside when the weather is unusually hot, in order to shield the human from the sun.

From the example above, it will be noted that our definition of "true" AI does not necessitate genuine human-level intelligence. Such a goal is still a long way from being achieved. In the meantime, AI systems may well give unusual outcomes that are not those a human would reach (although as we will develop further in this chapter, in some applications this apparent defect may be received with a "who cares?" response due to economic benefit outweighing human certainty). In our example above, the AI system, lacking as it does a full understanding of the motivating factors of human behaviour, would not be able to assess the fact that most humans in London would gain enjoyment from the sun, rather than wish to be shielded from it.

7.3 Possible applications of AI and machine learning in the maritime and trade world

There are too many potential practical applications of AI and machine learning with possible legal effect in the maritime and trade world to cover in one paper – or potentially one book.

Further, the processing technologies of AI and machine learning are not the only technologies likely to have a transformative impact on maritime and trade – the potential impact of developments in storage technology using blockchain are significant. For example, although the various applications of blockchain to the maritime world are the subject of other papers (and therefore beyond our scope) the potential uses are many and varied:

i) The use of blockchain to enable the widespread adoption of electronic bills of lading has been widely discussed.[3]
ii) Similarly, earlier this year Standard Chartered announced it had invested in Contour, which seeks to apply blockchain to letters of credit.
iii) Elsewhere, blockchain is used for real-time information sharing. Navozyme, based in Singapore, have developed a system called N-MAP allowing real-time authenticated information sharing between various stakeholders including shipowners, ports and classification societies.
iv) Blockchain is also used for tracing, including tracing bunkers through the supply chain (see the work undertaken by Blockshipping); Blockchain can also assist with the tracing of containers (see the Global Shipping Container Platform and ShipChain).

3 See the contributions of Professor Andrew Tettenborn and John Russell QC to this volume, Chapters 1 and 2.

Smart contracts will have numerous applications throughout the industry. A smart contract is, in essence, a computer algorithm that automates processes depending on inputs. So a smart contract operating in relation to a voyage charter may, for example, receive verification through blockchain of the quantity of cargo loaded on board the vessel (the bill of lading being itself electronic), and calculate the appropriate freight payment accordingly. This then allows the contract itself to facilitate and issue an instruction that the appropriate payment should be made in favour of the shipowner an appropriate number of days after the bills of lading date. Given agreement between the parties, more complicated scenarios could also be imagined. Imagine a shipbuilding contract. If the smart contract receives appropriate verification (agreed between the parties in advance) that a particular milestone has been met, it may give an automatic instruction that results in payment for that milestone being made by the Buyer to the Yard.

Data analytics will also drive change. TradeLens, a collaboration launched in 2018 by Maersk and IBM aims to bring together cargo owners, carriers, freight forwarders, logistics providers, ports, customs authorities and other such parties in a data gathering and information sharing platform to streamline container port operations. Although not necessarily involving AI or machine learning techniques, doubtless the application of machine learning principles to the data such a platform will provide will, in due course, yield great operational efficiencies.

Of more direct relevance, there is considerable current interest in using machine learning systems to optimise route planning. The Great Intelligence, a 38,800 dwt vessel with a smart navigation system that aims to be able to suggest a route based on a variety of preferences – for example, the quickest route, the most fuel-efficient route or the most "comfortable" route – was unveiled at the Marintec 2017 conference.[4] A collaboration between Stena Line and Hitachi using an "AI assistant" to minimise fuel consumption had reported fuel savings of 2–3%.[5] One can see that machine learning techniques, based upon large data sets of current and weather obtained by a number of vessels, will only increase such options – and that such options bring with them the possibility of a conflict between the interests of the various parties to the charterparty and bills of lading.

Navigation systems are also being improved through AI and, particularly, machine learning. Orca AI attempts to locate and track other vessels at sea, and suggest collision avoidance information. Navi-Planner generates voyage plans based upon the latest charts and environmental information. Such systems may well contribute significantly to vessel safety in the coming years.

We consider that the availability of insights gathered through the application of machine learning techniques to large datasets could itself lead to new regulatory obligations in the maritime sector, and perhaps from unexpected places. For instance, Windward have developed a system which tracks the movements of vessels, and analyses those movements to assess the risk that those vessels are engaging in questionable behaviour. This information can be made available not only to law enforcement, but to those contemplating dealing with those vessels.

4 "Smart Shipping, Moving Forward" (Lloyds Register, 07 December 2017) at www.lr.org/en-gb/latest-news/smart-shipping-moving-forward/ (accessed 1 May 2021).

5 "Efficiency: AI Assistant Installed on M/S Skåne" (Stena Line, 05 December 2019) at www.stenalinefreight.com/news/efficiency-ai-assistant-installed-on-m-s-skane/ (accessed 1 May 2021).

While there may be some in the industry who would simply prefer not to know if the vessel they wish to fix for a voyage appears to have previously been engaged in international sanction breaches (in much the same way that not every charterer followed the guidance in BIMCO's "Check before fixing"), once the tool is available, then regulatory authorities may well expect the industry to use it. Putting pressure on "innocent" shipowners and charterers to vet their counterparties in ways unimaginable until recently may well be a potent tool for the US Department of Treasury's Office of Foreign Assets Control (OFAC), which administers economic and trade sanctions – and if shipowners, why not hull or P&I insurers? A single sanctions-breaking voyage, detected by an algorithm comparing suspicious to non-suspicious behaviour, may leave a vessel untradeable in legitimate markets.

However, to demonstrate the impact that such technologies will have on maritime and trade law, we wish to consider the future of autonomous vessels.[6] We will consider two issues in particular relating to autonomous vessels:

i) First, what it will mean for an autonomous vessel to be "seaworthy" and how the owner will exercise due diligence to make it so; and
ii) Second, how the casualty investigation of a collision involving an autonomous vessel will differ from that of a manned vessel, and what sort of evidence may be gathered.

We have chosen these examples because the first is an example of how the introduction of AI will, we believe, throw up difficult problems for old legal concepts – and perhaps lead to what will appear to be odd results. We have chosen the second because we consider that it demonstrates how AI will in some cases lead to a reconsideration of what relevant evidence will be, and how it should be sought. Together, therefore, they give insight into the impact of AI on both the substantive and procedural elements of maritime and trade law. Of course, autonomous vessels will give rise to entirely new legal issues also. It is likely that there will have to be significant alteration of existing treaties, or perhaps, as Sir Bernard Eder has suggested,[7] a new overarching autonomous vessel code; furthermore, the increased threat of a cyberattack will raise fascinating issues relating to potential liability.[8] The latter is, however, the subject of another paper, and therefore outside the scope of this one.

7.3.1 Autonomous vessels (1): seaworthiness and due diligence

On 9 November 2018, Sir Bernard Eder gave the Inaugural Francesco Berlingeri Lecture for the Comite Maritime International,[9] on the subject of unmanned vessels. At [31], he

6 We assume that many of the current projects will be well known to those reading the paper. However, for those new to the subject, the Yara Birkeland, a fully automated containership, is in development in Norway. Until work was paused due to the COVID-19 pandemic, it was intended that the vessel would switch to a fully automated operation by 2022. In December 2018, Finferries and Rolls-Royce unveiled the FALCO, a ferry that navigated autonomously between Parainen and Nauvo, conducting automatic collision avoidance and berthing. In China, Qingdao shipyard has commenced construction of an autonomous container vessel, the ZHI FEI, due to be delivered in June 2021.

7 "Unmanned Vessels: Challenges Ahead" [2019] LMCLQ 47 at 55.

8 For a comprehensive analysis on these issues, see, B. Soyer and A. Tettenborn (ed), *Artificial Intelligence and Autonomous Shipping: Developing International Legal Framework* (Hart Publishing, 2021).

9 Comité Maritime International, *Inaugural Franceso Berlingieri Lecture, Unmanned Vessels@ Challenges Ahead* (Sir Bernard Eder, 09 November 2018) at https://files.essexcourt.com/wp-content/uploads/2018/11/08152752/Berlingieri-Lecture-FINAL.pdf (accessed 1 May 2021).

raised the subject of automation, seaworthiness and due diligence in a manner that bears quoting at length:

> First, a threshold question arises with regard to the potential legal liability of a shipowner in circumstances where, for example, an autonomous vessel is navigated from ashore and there is a collision or grounding as a result of a software problem caused by some third party – for example, the manufacturer or installer of the automation system or internet provider. In truth, this is not necessarily very different from the legal problems which can arise in the conventional context. In each case, the broad question arises as to whether the shipowner can avoid liability because of the fault of the manufacturer or installer of the software system or the third party provider. In the context of the Hague Rules, this in turn will focus on the scope of the obligation of due diligence to make the ship seaworthy before and at the beginning of the voyage under Art III.1; and the various defences which may be available under Art IV.2 including, of course, sub-paragraph (p) – "latent defects not discoverable by due diligence". In one sense, these are not new problems at all. However, as automation systems become more complex, one may assume that these issues will perhaps become increasingly important. Similarly, it seems to me that the question of rights of recourse will also become increasingly significant – and complex.

Sir Bernard was of course correct in his statement that these are not new issues. However, we consider it worthwhile considering how they may "play out" in the context of an autonomous vessel. Let us take Sir Bernard's example of a grounding caused by a software problem, and assume that the shipowner makes a claim in general average. This is rejected by cargo interests on the grounds of the vessel's unseaworthiness.

Art. III(1) of the Hague Rules reads in material part as follows:

> the carrier shall be bound before and at the beginning of the voyage to exercise due diligence to:
>
> (a) Make the ship seaworthy

The conventional test as to the meaning of unseaworthiness is "Would a prudent owner have required that it [the defect] should be made good before sending his ship to sea, had he known of it?"[10] Clearly, a prudent owner does not send a ship to sea with a navigation system likely to make it run aground, and therefore *prima facie* the vessel appears to be unseaworthy.[11]

Further, the fact that the relevant system fault will likely have been caused by the actions of a third-party software company is not itself a defence, as it is trite law that the shipowner's obligation to provide a seaworthy vessel is non-delegable. However, the shipowner's liability for defects predating the beginning of the voyage is not unlimited. In *The*

10 Originally from *Carver on Carriage by Sea*, approved in *McFadden v Blue Star Line* [1905] 1 K.B. 697 (Channell J.) at 706, and by the Court of Appeal in *F.C. Bradley v Federal Steam Navigation* (1926) 24 Ll. L. Rep. 446 at 454.

11 There are two slight caveats to this position. The first is factual. Given it may not be possible to interrogate the actions of an autonomous vessel to discover *why* it has acted as it did (as discussed in relation to collisions below), the actual diagnosis of a software fault may not be straightforward. However, we would argue that if physical issues can be ruled out as a potential cause, a software fault would be the only remaining diagnosis.

The second caveat is legal. The Court of Appeal has recently clearly stated that under English law an error of navigation prior to the commencement of the voyage (and we consider it arguable that a system fault could be considered such) can amount to unseaworthiness – *Alize 1954 v Allianz Elementar Versicherungs AG (The CMA CGM Libra)* [2020] EWCA Civ 293 at [70]. However, we understand that permission to appeal this matter to the Supreme Court has been given. Should the Supreme Court overturn the judgments of the Court of Appeal and Admiralty Court in this regard, an extra layer of complication could be added to the analysis.

Muncaster Castle,[12] the House of Lords drew a distinction between unseaworthiness that occurred as a result of actions during the period of the shipowner's control of the vessel and those predating such control in the following terms:[13]

> The carrier's responsibility for the work itself does not begin until the ship comes into his orbit and it begins then as a responsibility to make sure by careful and skilled inspection that what he is taking into his service is in fit condition for the purpose and, if there is anything lacking that is fairly discoverable, to put it right. This is recognized in the judgment. But if the bad work that has been done is "concealed" (pp. 462 and 214 of the respective reports) and so cannot be detected by any reasonable care then the lack of diligence to which unseaworthiness is due is not to be attributed to the carrier.

One would at first blush expect that the software fault would be the result of coding during construction of the vessel, and therefore, by this definition, fall outside of the ambit of the Owners' responsibility,[14] so long as the shipowner can establish that it exercised due diligence to discover the defect.[15]

However, consider the fact that the automated vessel will be engaged in machine learning through experience, using the processes set out in Part II.[16] The decisions that the vessel takes in any given situation will be bound up in the way in which it has interpreted this experience. This means that it may well be impossible to ever tell whether the software flaw that caused the grounding in our example arose before or after delivery of the vessel, and therefore whether the vessel became unseaworthy while within the shipowner's orbit or not. If so, then the position would be reversed, and it is difficult to see how the shipowner could ever escape liability for an unseaworthy vessel on the basis that the unseaworthiness pre-dated the vessel entering the shipowner's orbit.

The question of whether the shipowner has exercised due diligence to make the vessel seaworthy then arises, which in turn raises the question of what a shipowner can do to discover a defect in code with which he is presented. Is the shipowner expected to hire a further developer to undertake its own verification of each and every line of what will be an undoubtedly complicated system and which, following some period of machine learning, may not be readily comprehensible to any human in any event?[17] This would appear unlikely to be reasonable. There may be some diagnostic checks available on the software

12 *Riverstone Meat Co Pty v Lancashire Shipping Co (The Muncaster Castle)* [1961] 1 Lloyd's Rep. 57.

13 Ibid., at 85, commenting on *Angliss & Co. (Australia) Pty., Ltd. v Peninsular and Oriental Steam Navigation Company* [1927] 2 K.B. 456, (1927) 28 Ll.L.Rep. 202.

14 See, for example, *Parsons Corp v CV Scheepvaartonderneming Happy Ranger (The Happy Ranger)* [2006] EWHC 122 (Comm); [2006] 1 Lloyd's Rep. 649, in which Gloster J. held that the Owners' "*orbit*" should be construed "co-extensively with ownership or service of control" (at [2006] 1 Lloyd's Rep. 649 at 656). A claim that the shipowner should be considered liable for a defect that occurred during the construction of the vessel where the shipowner had been closely involved with the construction process was rejected, as the relevant shipbuilding contract made clear that the shipowner did not assume responsibility for the vessel until after delivery (although the shipowner was held to be liable on other grounds).

15 That the burden of proof is on the shipowner to establish that due diligence has been exercised for the purpose of Article III(1) was assumed to be true by Gloster J. in *The Happy Ranger* at 658, and was confirmed by Teare J in *Alize 1954 v Allianz Elementar Versicherungs AG (The CMA CGM Libra)* [2019] EWHC 481 (Admlty); [2019] 1 Lloyd's Rep. 595 at 605. An appeal on this point by cargo interests was not pursued (see the Court of Appeal decision, *Alize 1954 v Allianz Elementar Versicherungs AG (The CMA CGM Libra)* [2020] EWCA Civ 293; [2020] 2 All E.R. (Comm) 1072 [12]).

16 Indeed, it is likely that the vessel will be connected to a number of vessels, the experience of all of which will be relevant in determining how any individual vessel reacts to a given situation.

17 Indeed, it is possible that those who created the code may seek to keep it confidential in any event.

that it would be expected the shipowner will run occasionally. Assuming these checks have been made, then it becomes possible that *any* flaw in the coding of an autonomous vessel does not render that vessel unseaworthy.

Furthermore, an autonomous vessel will not only have an autonomous navigation system, but also an autonomous cargo handling system. Assume now that a fault in such a system leads to a claim under a bill of lading in relation to a breach of the carrier's obligation to "properly and carefully load, handle, stow, carry, keep, care for, and discharge the goods carried", and/or under a charterparty in which the Hague Rules have been incorporated. If the vessel is not unseaworthy, the shipowner will be able to rely upon its article IV defences, including (as Sir Bernard noted), article IV(2)(p) – "latent defects not discoverable by due diligence". Following the analysis above, it is likely that any such system failure would constitute a "latent defect".

While, therefore, it is hoped that one advantage of automation will be an improvement in vessel operations and safety, it is possible that the mere *fact* of automation will provide a shipowner with additional defences where an incident does arise.

7.3.2 Autonomous vessels (2): collisions

There has been some attention already given to the question of whether an autonomous vessel would be inherently unable to comply with the International Regulations for the Prevention of Collisions at Sea (COLREGS).[18] We do not intend to consider this position in this chapter. Instead, we wish to consider something more basic – how will a future casualty investigation approach a collision between an autonomous and a manned vessel?

As a scenario, let us assume that there has been a collision between an autonomous vessel and a manned vessel. There is no dispute that the autonomous vessel was the give-way vessel in a crossing situation where there was a risk of collision, but it failed to take any substantial action to keep well clear of the (manned) stand-on vessel until a collision was entirely inevitable.

For those acting on behalf of the manned vessel, it is likely that little will change. The investigator will save the VDR data, working chart, vessel logs, course recorder trace and other similar documentation, and proof the crew in an entirely orthodox manner. The judge can consider this contemporaneous evidence alongside witness evidence to show what such evidence reveals about (for example) the situational awareness of those on board the

18 See, for example, Comité Maritime International, *Inaugural Franceso Berlingieri Lecture, Unmanned Vessels: Challenges Ahead* (Sir Bernard Eder, 09 November 2018) [43(b)]; and Danish Maritime Authority, *Analysis of Regulatory Barriers to the Use of Autonomous Ships,* (December 2017) para 3.4.1 at www.dma.dk/Documents/Publikationer/Analysis%20of%20Regulatory%20Barriers%20to%20the%20Use%20of%20Autonomous%20Ships.pdf (accessed 1 May 2021).

Both conclude that there would be severe difficulties in compliance, not only with the Colregs, but indeed with a wide variety of currently enforceable treaties operating in the maritime world. The presence of such difficulties is one reason why we do not below discuss the application of the Colregs to autonomous vehicles – it seems likely to us that the widespread adoption of autonomous vehicles will necessitate the introduction of a new set of collision rules for autonomous vehicles. While the underlying substance of the rules is unlikely to change (e.g. for example, the definitions of a stand-on and give way vessel will remain), such rules will be drafted in a manner which recognises the crewless nature of the vessel. As such, it may well be the case that in a future collision between an autonomous vessel and a manned vessel, the actions of the manned vessel will be judged against the standards of the Colregs, whereas those of the autonomous vessel will be judged against the autonomous equivalent.

vessel, the standard of the watch kept (whether visual or aural), and the extent to which the reactions of those on the bridge to a developing risk of collision appeared to be taken on a rational basis, and in a timely fashion.

For the autonomous vessel, it is likely that those investigating the collision (and the eventual judge) will have access to a great deal of automatically captured data – and quite possibly data that is more extensive, and more accurately recorded, than for a non-autonomous vessel. After all, sufficiently accurate sensors to allow a detailed real-time understanding of the environment in which it operates will be a necessary precondition of any level of automation. It would further be expected that an autonomous system would keep a precise time-stamped record of every navigational instruction, whether related to engine speed or course.

This material may reveal the collision had an immediate technical cause. It may be, for example, that the vessel keeps logs of the objects it considers it has identified, and there is no record of identification of the stand-on vessel at all. By way of analogy, in March 2018 an Uber self-driving car tragically struck and killed a pedestrian crossing the road. The subsequent report of the US National Transportation Safety Board (NTSB) revealed that at no point before the collision had the vehicle's sensors properly identified the pedestrian or predicted her path. Only when the pedestrian was immediately in front of the vehicle, 1.2 seconds before the collision, did the vehicle's system recognise an emergency breaking situation had occurred. It appears this may have been because the vehicle's collision detection system did not recognise the possibility of jaywalking pedestrians.[19] However, there can be no guarantee that the immediate cause of the collision will be so clear.

We, therefore, suggest that the first line of investigation on behalf of an autonomous vessel may well not be carried out by a standard mariner investigator, but instead by a software engineer who has maritime expertise. Where the cause is not immediately obvious, the first step may be to run simulations of the incident using the data gathered from the vessel and see whether, in the simulation, the collision still occurs. A system given the same inputs should produce the same outputs each time.

It may be, for example, that in the simulation the autonomous vessel takes early and substantial action to avoid the stand-on vessel that was not taken in real life. This would suggest that the autonomous vessel had recognised the risk of a collision, and had correctly calculated the best way to avoid it, and had attempted to put that calculation into practice. Investigations can then consider why, in fact, the collision still occurred. For instance, there may have been a mechanical breakdown in the steering gear, such that the "orders" to change course given by the vessel's autonomous systems could not be effected. Alternatively, it may be that the algorithms concerning navigational direction and the operation of the steering gear did not correctly communicate, such that the "order" to turn was not effected, even though the steering gear was physically operational. It may even plausibly be the case that the sensors relating to the direction and strength of current and/or wind were faulty, leading to the vessel significantly underestimating the magnitude of the course correction needed.

19 National Transportation Safety Board, *Collision Between Vehicle Controlled by Developmental Automated Driving System and Pedestrian, Tempe, Arizona, March 18, 2018* (Highway Accident Report NTSB/HAR-19/03, 2019) para 5.6.1 at www.ntsb.gov/investigations/AccidentReports/Reports/HAR1903.pdf (accessed 1 May 2021).

On the other hand, it may be that the simulation shows the collision occurring again. At this point, the benefit of a simulation would be that various parameters can be changed so as to try and identify the fault. For instance, if collision avoidance occurs when the size of the other vessel is increased, night is turned to day (or vice versa), or the navigational lights displayed by the other vessel are made more prominent, it would seem probable that the fault would lie with the vessel's visual recognition software, in some form or other.

The question will then arise as to how such information should be dealt with by the Court. In this regard, it is likely that even an entirely autonomous vessel would have a remote operator with overall responsibility for the vessel, and the ability to take control if they considered that the vessel's automated navigation system was failing in some way. If so, identifying the cause of a collision is likely to be more complicated than simply finding an immediate technical cause for the collision.

In this regard, the NTSB's report into the Uber self-driving vehicle is again instructive. Despite the technical failures described above, the report's determination was that:

> the probable cause of the crash in Tempe, Arizona, was the failure of the vehicle operator to monitor the driving environment and the operation of the automated driving system because she was visually distracted throughout the trip by her personal cell phone. Contributing to the crash were the Uber Advanced Technologies Group's (1) inadequate safety risk assessment procedures, (2) ineffective oversight of vehicle operators, and (3) lack of adequate mechanisms for addressing operators' automation complacency – all a lack of proper operational procedures and an inadequate safety culture. Further factors contributing to the crash were (1) the impaired pedestrian's crossing of N. Mill Avenue outside a crosswalk, and (2) the Arizona Department of Transportation's insufficient oversight of automated vehicle testing.[20]

In other words, despite the immediate technical difficulties the vehicle experienced, the NTSB's report focussed on a variety of human failings, including those of the vehicle operator, of Uber, of the pedestrian and the Arizona Department of Transportation.

Returning to a collision between an autonomous and manned vessel, consider the following potential scenarios:

i) Imagine firstly that an operator has noticed that the vessel's sensors do not appear to be working correctly, and the vessel must be watched carefully. As their shift ends, their replacement is running late due to traffic in Mumbai. As a result, the handover is rushed, and the replacement operator does not fully comprehend the issue with the vessel's sensors, and does not pay requisite care. Is this a human or a systems error, or simply two separate faults?

ii) Alternatively, imagine an operator tasked with operating multiple autonomous vessels at the same time. They are distracted by an issue on another vessel, and simply fail to notice the situation of danger developing on the subject vessel because of a previously unnoticed sensor error. Do similar considerations apply?

iii) Another alternative could involve the apportionment of responsibility where there is both a defect in the system but also evidence of fatigue/burn out due to the operator being asked to work hours in excess of their contract, or any applicable

20 Ibid., at v–vi.

maritime labour convention working hours limit as a result of staff shortages due to a pandemic. Here there may be fault on the part of the system, the programming, the operator and the employer.

The essential distinction between these scenarios may be said to be that in the first, the technical fault is known but not sufficiently comprehended; in the second, the technical fault is unknown, but discoverable had greater diligence been exercised. The third is of course a combination of numerous causative factors. How a Court would consider the interplay between the technical and human errors in each case is, we consider, somewhat difficult.

Further, imagine a diligent operator has noticed what appears to be a situation of danger developing. As the vessels appear to converge and a clear risk of collision develops, the operator becomes increasingly concerned. How should that operator react?

One of the major advantages of autonomous vehicles is said to be safety, in that an autonomous system will likely take better decisions that a human would. Assuming this is so, and assuming there is no obvious and known fault with the vessel, there may well be a tendency on the part of a human operator to place too much trust in an autonomous vessel's own collision avoidance capabilities, and take no action until it is too late. How would a court view the failure by the operator to take any action, and effectively watch the collision occur? Does watching a collision occur and doing nothing in the hope that an automated system has the correct strategy carry greater causative potency than a *failure* of lookout on a manned vessel?

By way of contrast, imagine the operator panics, and takes control of the vessel. In taking the wrong action, the operator causes the very collision they were seeking to avoid. Subsequent simulations show that the engine orders given by the vessel's AI navigation system would have avoided the collision. Should the operator bear *greater* responsibility than the Master of a manned vessel who made the wrong decision in a similar situation, given that the operator did not need to take any action, and were it not for their actions, no collision would have occurred at all?

In summary, we consider the interface of human and technical factors will likely cause new difficulties in assessing liability for collisions – maritime lawyers are likely to puzzle more, rather than less, over such questions in the future. However, the position is no simpler where there is no suggestion of a failure by a human operator, but a simple systems failure, either of programming or sensors. Can one equate a fault of visual recognition software with a defective human lookout – is a computer bug similar to a failure by the Master of the manned vessel to look out of the bridge window?

This raises the question of how a human judge would weigh the severity of such mistakes – would there be a tendency to consider human failings to be more culpable than mechanical ones in a situation where VDR audio data and a self-justifying witness statement may vividly bring home human failings in a way that the autonomous vehicle can avoid?

The judge will be in the difficult position of having the usual access to all of the evidence regarding the decision-making process on board the manned vessel, and will have to weigh that against a system which does not make "decisions" as such at all, and/or in which there will be a complicated interplay of faults of man and machine that will likely be hard to untangle.

In such a situation, we would hesitantly suggest that the interests of justice may lie in simply discarding much of the usual evidence on which the judge would rely, and simply recreating the speeds and courses of both vessels using such data as is available, and attempting to reach a conclusion based on that alone. Otherwise, while the mistakes, misunderstandings and miscommunications of the human crew or operator will be readily comprehensible, the flaws in the system of the autonomous vessel, being outside of human terms of reference, will be comparatively opaque; and it is not too fanciful to suggest that in such a situation, the actions of the human crew will be judged more harshly than those of the autonomous vessel.

In any event, however, it is likely that collision investigations involving autonomous vessels will change fundamentally. The investigative element will not focus merely on the physical vessel and its crew, but will likely involve recreating the collision electronically. The relevant skills will involve assessing which parameters in the recreated collision need to change in order to prevent the collision occurring, and extrapolating from those changed parameters to a theory of why the collision occurred. In these circumstances, we would suggest that if the underlying causes of a collision are to be considered in Court, it may be necessary for a lay judge to be assisted not only by nautical assessors in uniform, but technical assessors in black T-shirts.

One final risk area which is again beyond the scope of this chapter, but is worth drawing attention to, is the role of the hacker or maritime terrorist in future casualty scenarios. Where the incident has occurred due to a breach of cyber security or an attack on operational technology, where will the fault lie? Moreover, in the future as the sophistication of cyber attacks increases, what if the attack is as a result of autonomous malware and adversarial machine learning – who bears legal responsibility where the source of an attack is a hivenet or swarmbot – and will this even be discoverable as it is expected such systems will self-delete after penetration.[21]

7.4 Changes in legal practice

The question considered in this section can be shortly put: if the substantive and procedural elements of maritime and trade law will be heavily affected by the introduction of AI, will there also be fundamental changes in how maritime and trade law services are provided?

We have no doubt that, as with any topic gaining attention at any particular time, terms such as "AI" and "machine learning" will be used by a variety of snake oil vendors. We suspect there are a number of players in the market seeking to dress up fairly standard algorithms (including poorly-designed algorithms) as "AI" in the hope of increased marketing appeal. This does not mean that "true" machine learning and AI systems cannot be transformative, however.

Certainly, the introduction of AI is currently providing, and will continue to provide, avenues by which such services can be provided more efficiently. We have already mentioned

21 Cyber security risks are the subject of another paper, and therefore will not be covered in any detail here. Briefly however, "adversarial" machine learning seeks to overcome a defensive system such as by having the system misclassify at attack through the use of misleading inputs. In order to create such inputs, the "attacking" system will need to be able to predict how the "defensive" system may respond to an attack, so as to create an input that can mislead the defensive system. Although at present there is no entirely autonomous malware, the future development of hivenet and swarmbots has been theorised. In these systems, the individual bots are expected to be able to communicate with each other and the "hive" thereby allowing a large number of autonomous bots to share information and common strategy.

above the use of AI in document review software, which allows very significant savings to be made in time, legal fees (and the sanity of junior lawyers) in the reviewing of large volumes of documents.

Similarly, contract automation services that allow parties to choose from a menu of clause "options" for standard form contracts have been available for some time. Such services are perhaps slightly different from the litigation support software in that, after the initial legal input of drafting the various clauses has taken place, may allow a sophisticated client to dispense with the lawyer altogether. However, it may perhaps be doubtful whether an automated contract alone qualifies as "AI". More interestingly, we discussed smart contracts above. One can imagine that such contracts will of necessity lead to a great deal of standardisation, and the lead on their creation will be taken by programmers rather than lawyers. We know of at least one leading smart contract legal specialist that has embarked on training as a programmer and writer of legal code. We suspect this kind of expertise will become common place in the future.

There are also already systems available which claim to be able to perform their own drafting. IBM's Watson system claims to be able to automatically draft simple responses to initial complaints in US litigation, and formulate initial discovery requests. Walmart apparently has licensed the system for use in all slip-and-fall lawsuits filed in California.[22] It would be naïve not to expect that similar systems will be introduced in due course in the UK. Such systems at first are likely to be for high volume, uncomplicated slip-and-fall type claims, but one can see how increasing sophistication and comfort with their use will lead to their adoption more broadly – including, likely, in the world of maritime and trade law, especially in the field of small claims.

The processes described above are likely essentially to increase efficiency by standardising and mechanising existing processes. In doing so, they will likely lead to a net decline in the number of matters needing individual input from a lawyer. However, each of the above systems will still need legal input in the design and implementation phase. A programmer will find a way to ensure that a smart contract recognises that after event X happens, the process Y rather than process Z should be executed. However, legal input will likely be necessary before the coding occurs to consider when Y or Z should be the appropriate response. The more complicated the contract, the less trivial such a decision is likely to be. This will be legal work of a very different type and requiring a different mindset to traditional contract drafting. It will fundamentally change the preconception of what a lawyer does all day.

Lawyers are also likely to become more engaged with creating and managing the regulatory framework in which our ever advancing AI and machine learning systems operate. Who will govern cyber space and protect us against a potential "Skynet" event? Pure science fiction or justifiable cause for concern?

In addition to increasing efficiency, AI is also introducing entirely new products to legal services. For example, Lex Machina, now part of LexisNexis, offers a "legal analytics" system based upon an automated review of online US legal information, and claims to be able to provide guidance on a number of case-relevant issues. For example, the system will consider specific judges, including how long the judge will likely take to reply to a motion

22 T. Suh and J. M. Lee, "Save the Lawyer: AI Technology Accelerates and Augments Legal Work" (*IBM*, 07 August 2018) at www.ibm.com/blogs/client-voices/save-the-lawyer-ai-technology-accelerates-and-augments-legal-work/ (accessed 1 May 2021).

based on previous experience, and even how likely they are to grant a motion of a particular type. It will consider the experience of individual opposing counsel, and assess how experienced they are in the type of work; it will review the experience of the counterparty, and what their experience has been in the courts previously. This is done through the analysis of documents filed in 2.8 million civil cases in federal courts, and 1.3 million cases from nine state courts.[23]

Although such an approach would be more difficult to copy in the world of English maritime and trade law given the considerably smaller number of cases on which a system may be trained, we consider it likely that the influence of data analytics on such cases as are available will be a fruitful future area of research. Again, legal input will be needed not in the manner of traditional legal work, but in order to imagine what insights may be useful for the lawyers, so that those with appropriate skills can then design the program to find them. Again, legal work changes rather than disappears.

We, therefore, take it as a given that AI and machine learning will be driving changes to legal processes in the future. The question then becomes where the boundaries for such changes may lie. In the litigation sphere, we see the increased use of automated systems for resolving straightforward insurance claims, and apply the same logic to demurrage, speed and performance, routine cargo damage and shortage claims, personal injury and small claims. The bigger question is whether such utilisation of AI will remove lawyers from the equation in some areas altogether. We suspect that it cannot (or at least, not yet), and that we are some way from a position where AI alone can resolve a "dispute", rather than a straightforward claim. Given there have been some reports in the press concerning "robot mediators" we also briefly consider mediation – again, for anything other than the most straightforward of claims, we doubt that AI will replace humans quite yet.

7.4.1 Insurance claims automation

In the insurance industry, there is already a keen interest in the greater use of technology to determine, assess and pay claims. At the consumer level, Ageas recently claimed to have been the first insurance entity to use AI to create car damage assessments and repair assessments entirely remotely.[24] When a car is damaged, the customer will upload photographs from their smartphone to a website where AI (presumably a system utilising image recognition through machine learning) will be used to provide a full estimate of cost, including work to be completed and labour hours.

Similarly, Lloyd's has published its Blueprint Two, a comprehensive plan for the increased use of technology across the insurance market. This includes the management of claims; it is intended that:

> We will deliver a platform that will lead to a seamless customer experience designed to meet their expectations. Straightforward claims will be resolved automatically (or with limited

23 G. Huang, "Lex Machina Announces Expanded Federal District Court Coverage" (*Lex Machina Blog*, 05 August 2020) at https://lexmachina.com/lex-machina-announces-expanded-federal-district-court-coverage/ (accessed 1 May 2021).

24 "Ageas is first UK insurer to use AI to create end-to-end car damage assessments and estimates" (Ageas, 26 March 2020) at www.ageas.co.uk/press-releases/2020-press-releases/ageas-is-first-uk-insurer-to-use-ai-to-create-end-to-end-car-damage-assessments-and-estimates/ (accessed 1 May 2021).

> manual touch points). More complex claims will be supported by greater use of collaboration and workflow solutions. All claims will be supported by better data driven from placement as well as by a suite of centralised support functions.[25]

It may be striking for many that it is envisaged that the *entirety* of the claims handling process for simple claims may be automated. It may perhaps be thought that (in a similar way to some of the other claim types detailed above) such a system will, simply, sometimes be entirely wrong, and reach decisions that a human adjuster would not reach, in terms of quantum or the interpretation of clauses. However, broadly speaking, it is easy for lawyers to forget that most claims in the insurance industry are paid, broadly in the amounts claimed. If the system is occasionally inaccurate in a way that advantages the customer, this will be to the detriment of the insurer in the specific case – but at the same time, the insurer will have avoided the transaction costs of a whole basket of claims which are paid without issue. Economics outweighing legal certainty.

Lawyers looking at the outcome of an algorithm's decision in a particular case may say algorithms get things wrong. We only need to look at the recent example in England and Wales concerning the evaluation of A-level examination results. But in an ever-increasing technological world right and wrong may be surpassed by commercial generality. Are we approaching a position where so long as a system gets it right most of the time, we will be content that due to the saving in cost this new ambiguity in legal certainty outweighs absolute correctness of the decision-making process?

Against this background, it would be surprising if increased automation did not affect claims processing in marine insurance lines, as well as the market as a whole. It is most likely that claims up to a particular value will be entirely automated, through interaction with an online portal which will take details of the claim, and through which relevant supporting documents will be uploaded. The system will then undertake a simple assessment as to whether it appears the loss was caused by an insured peril, and make a recommendation for payment. The savings will come not only from a reduction in claims handling time and, therefore, staff (although there will likely be occasional random human review as a deterrent to outright fraud), but also in saved third-party costs – it may no longer be automatically the case that a surveyor would be sent to view the vessel, for example this being replaced by the use of drone technology and remote investigation.

Although complicated, higher value claims will still require human input, one can see that this will inevitably lead to somewhat of a decline in the number of insurance cases reaching the desks of lawyers. At the same time though, this opens up the possibility of a new stream of work for lawyers in assisting with the design of automated systems when determining a claim. In other words, lawyers would not just be selling their skills in resolving individual claims, but also their abilities to create an approach to resolving claims that have not yet occurred.

7.4.2 Demurrage and automation

The question then becomes whether, if an insurance claim can be automated, can another form of "dispute" be resolved through AI? To us, demurrage claims appear to be an obvious

25 Lloyd's, "The Future at Lloyd's: Blueprint Two" (5 November 2020) at https://futureat.lloyds.com/lloyds-blueprint-two.pdf (accessed 1 May 2021).

contender for such resolution; in essence, calculating the correct amount of demurrage owed should be a largely arithmetic endeavour, based on factual circumstances that are (mostly) objectively identifiable. Furthermore, there is a highly developed body of law relating to the interpretation of laytime and demurrage clauses on which a computer may be trained to give relevance to such factual circumstances. We, therefore, use demurrage below as a stand-in, in some respects, for other types of claim to which similar principles may apply.[26]

The calculation of demurrage relies upon statements of fact from ship and shore, notices of readiness, pumping logs and letters of protest, and so on. Such documents are prepared by humans, and are therefore subject to all of the inaccuracies which human intervention can bring, including errors in recording data, and omissions through oversight of recording relevant events. We do not envisage such documents will be relied upon in the future. As a necessary step towards automation, it will be the case that sophisticated sensors must be installed upon vessels – the feedback such sensors give will be a necessary part in enabling any automated vessel control system to direct the actions of a vessel. Similar developments are also envisaged on the shore side[27] to create smart ports.

The effect of these changes will be to create a stream of real-time, electronically generated data from two independent sources, verified through blockchain, which can be used to verify the precise time and location at which almost any step in the loading and discharging process occurred. For example, an electronic notice of readiness can bear the vessel's coordinates at the precise time of submission. The precise time of free pratique (given electronically) can be placed into an electronic statement of facts. The times at which hatches are opened and closed, and at which hoses are connected and disconnected will be clear. When loading stops due to rain this is something that can be determined with exactitude. Pumping logs will be automatically generated, and so on.

It, therefore, takes little imagination to see how the recording of such streams of data could be used to generate an automatic laytime/demurrage calculation for the majority of shipments in which there are no issues in dispute. Indeed, in a majority of cases such a calculation could be generated largely in real time, and probably emailed to the intended recipient with printouts of supporting data from both ship and shore before the vessel has even dropped the last outward pilot.

Meanwhile, as is well known, certain voyage charters (including standard form charters, particularly in the tanker sector) impose strict requirements on Owners to provide such documents (or a subset of them) within a particular period, or waive their right to claim demurrage. A rich body of case law has developed purely in order to analyse such clauses, and parse the requirements that they place on Owners.[28] Where the data relied upon is

26 In particular, it would appear that disputes concerning the assessment of speed and performance, the ongoing calculation of hire and perhaps simple cargo claims may be subject to similar principles.

27 See Department for Transport, *Maritime 2050 – Navigating the Future* (January 2019) at para 6.4.

28 See, by way of example, *Tricon Energy Ltd v MTM Trading LLC (The MTM Hong Kong)* [2020] EWHC 700 (Comm); [2020] 2 All E.R. (Comm) 543; *Tankreederei GmbH & Co KG v Marubeni Corp (The Amalie Essberger)* [2019] EWHC 3402 (Comm); [2020] 1 Lloyd's Rep. 393; *Kassiopi Maritime Co Ltd v FAL Shipping Co Ltd (The Adventure)* [2015] EWHC 318 (Comm); [2015] 1 Lloyd's Rep. 473; *National Shipping Co of Saudi Arabia v BP Oil Supply Co (The Abqaiq)* [2011] EWCA Civ 1127; [2012] 1 Lloyd's Rep 18; *AET Inc Ltd v Arcadia Petroleum Ltd (The Eagle Valencia)* [2010] EWCA Civ 713; [2010] 2 Lloyd's Rep 257; *Petroleum Oil & Gas Corp of South Africa (Pty) Ltd v FR8 Singapore PTE Ltd (The Eternity)* [2008] EWHC 2480 (Comm); [2009]

automated, however, such requirements would be essentially obsolete – a clear advantage for the shipowner (albeit not for the charterer).

However, more interesting questions arise as to the ability of a computer program to determine an actual demurrage *dispute*. Let us imagine that a computer program operating on the basis of AI or machine learning has "read" every reported demurrage case and arbitration report in English law, and used them to develop a set of rules governing how demurrage calculations can be performed, in line with the principles set out in Part II. Such a system will still be subject to the following constraints:

i) The system will not overturn precedent. It has "learned" the rules, and will seek to apply them. The system will not recognise that a case has been wrongly decided, as it has no basis on which to do so. The law will, therefore, not evolve. However, we would note that while this is likely to be of concern to academic commentators and disappointing for practitioners, most clients may not consider such a state of affairs to be an inherent problem, if disputes could be resolved efficiently and cheaply.
ii) Related to (i) above, to the extent that there are conflicting decisions on a point, the machine learning system will likely be unable to provide an answer. It will have no way of favouring one decision over another. This would be a clear problem; in such a case the system would, presumably, have no way of reaching a decision at all.[29]
iii) The system will be able to reach a decision only in situations where it considers there is a rule that can be applied. Where a point is entirely novel, it will be unable to reach an answer. The pool of cases on which the system might train itself is relatively small.[30]

Assume, however, that for a given case, the rules to apply are sufficiently clear to the system. Let us set a scenario. Imagine that time at a port has been incurred due to a failure of plant in the port. The charterparty contains a clause akin to Asbatankvoy clause 8, which reads in material part (emphasis added):

> Charterer shall pay demurrage per running hour and pro rata for a part thereof at the rate specified in Part I for all time that loading and discharging and used laytime as elsewhere herein provided exceeds the allowed laytime elsewhere herein specified. If, however, delays occur and/or demurrage shall be incurred at ports of loading and/or discharge by reason of fire, explosion, storm, or by a strike, lockout, stoppage or restraint of labour, *or by a breakdown of machinery or equipment in or about the plant of the charterer, supplier, shipper or consignee of the cargo*, such delays shall count as half laytime or, if on demurrage, the rate of demurrage shall be reduced one half of the amount stated in Part I.

The first thing to note is that the application of the clause will depend on the identification of whether the breakdown occurred in the plant of the charterer, supplier, shipper or

1 Lloyd's Rep 107; *Waterfront Shipping Co Ltd v Trafigura AG (The Sabrewing)* [2007] EWHC 2482 (Comm); [2008] 1 Lloyd's Rep 286.

29 See, for example, the recent decision of Mr Justice Andrew Baker in *The Eternal Bliss* [2020] EWHC 2373 (Comm), in which the Judge considered the nature of precedent as it related to differing first instance decisions, and further whether differing first instance decisions actually applied to the point which had to be decided [52–54]. We do not consider it likely that the system in question would be able to undertake such an exercise.

30 J. Schofield, *Laytime & Demurrage* (7th edn), (Informa Law, 2016) contains a list of cases extending over 31 pages, with up to 50 cases on a page. Although this is a formidable work of scholarship, such a case list is unlikely to cover every question of legal interpretation that may arise. This is particularly so as many of the cases may (for example) be older cases or lower court decisions that are subsequently overruled.

consignee of the cargo. This is not a matter in which the sensors on board the vessel would be able to provide any insight. It will, therefore, be necessary for the system to take into account facts that are provided to it through some form of human intervention.

Second, assume our system has trained itself on all known reported demurrage decisions. It is, therefore, aware of *The Afrapearl*,[31] in which discharge at a port was delayed by a leaking sealine. Before the Court of Appeal, there was no dispute that the sealine was in or about the plant of the charterers, supplier, shipper or consignee of the cargo. The Court of Appeal held that a leaking pipe constituted a breakdown of machinery or equipment, as:[32]

> a breakdown of equipment such as the discharge pipe occurs when it no longer functions as a pipe. The cause of the breakdown may be a hole in the pipe or, as here, a gap in way of the flange which prevents the pipe operating as a discharge pipe. The hole may of course be caused in a number of different ways and for a number of different reasons.

Our system is also aware of *The Ladytramp*,[33] in which the Court of Appeal approved the earlier (unreported) case of *The Thanassis A*, in which it was held that the complete destruction of an object is different in kind to a "breakdown". Our system may, therefore, determine that in order to determine the correct sum of demurrage due, it must decide whether a piece of equipment is no longer functioning in the way that such equipment should, but nonetheless has not been completely destroyed. There may be times when such a fine distinction would be difficult for a human to decide; however, it is difficult to imagine the basis on which the system would resolve this at all.

It is, therefore, difficult to see a situation in which our machine learning trained system would be able to resolve such a dispute. We see the potential for sensors and smart contracts to provide an efficient initial demurrage calculation where there are no relevant issues that are extraneous to the physical experience of the vessel – but we doubt that such a system would be in a position to resolve a fully-fledged "dispute", or will be for some time yet.

7.4.3 Mediation

Last year it was reported that AI had even penetrated that most apparently human of areas, mediation, as a court dispute was apparently settled for the first time by a "robot mediator".[34] However, we would respectfully suggest that this somewhat overplays the technological sophistication of what occurred. The "robot" in question appears to have been a computer algorithm operating a blind-bidding system, that is, one in which both parties make high and low settlement bids unknown to each other. Where there is overlap, settlement will be taken to have occurred at the midpoint of the difference between the bids. If there is no settlement, further rounds of bidding can take place.

We do not consider this to be particularly technically sophisticated, and doubt that such a system would even fall within a broad definition of "AI". The makers of the system

31 *Portolana Compania Naviera Ltd v Vitol SA Inc (The Afrapearl)* [2004] EWCA Civ 864; [2004] 2 Lloyd's Rep. 305.

32 Ibid., at [21].

33 *ED&F Man Sugar Ltd v Unicargo Transportgesellschaft GmbH (The Ladytramp)* [2013] EWCA Civ 1449; [2014] 1 Lloyd's Rep. 412.

34 K. Beioley "Robots and AI Threaten to Mediate Disputes Better Than Lawyers" (*Financial Times*, 14 August 2019) at www.ft.com/content/187525d2-9e6e-11e9-9c06-a4640c9feebb (accessed 1 May 2021).

themselves note that it is "designed for two parties with simple negotiations that can be reduced to a single numerical issue".[35] Taken literally, this sentence would cover the majority of two-party mediations in the maritime and trade world: it is usually acknowledged between the parties (at least tacitly) that one party will be the payor, and the other the payee, and the only question is how much. However, those with experience of mediation will know that the key skill of a mediator is often in bringing home to the parties the advantage of settling the dispute immediately. We are not aware of any system that purports to be able to operate as a "human" mediator in bringing parties together, and nor do we envisage one being created any time soon.

Instead, where this system may have a future is in low-level disputes where the use of a human mediator would be entirely non-cost-effective. It appears that the court dispute resolved by the "robot mediator" was a Money Claims Online dispute over £2,000.[36] Such a system may indeed be helpful in certain cases, but we suggest that it is unlikely to garner widespread use in the resolution of complex disputes.

However, this is not to say that the "robot mediator" is the only way in which machine learning techniques may be applied to settlement discussions. One area in which strides have been made in recent years is in using machine learning techniques – and particularly the "random forest" technique referred to in Part II – as a way of predicting the outcome of certain cases, but without reaching the level of accuracy and/or detail that would be necessary to rely on the outcome of such cases for determination purposes.

By way of example, one system has been developed that has analysed every case that has been decided in the US Supreme Court.[37] The system claims to be able to predict the outcome of any case over that time period with 70.2% accuracy overall, and 71.9% accuracy at the level of any individual justice. It is easy to imagine such a system being helpful in settlement discussions. The system becomes a tool the parties can use as evidence of what may happen were a human judge to look at the case, and therefore guide and prompt the parties' negotiations.

However, the abilities of such a system should not be overestimated. As regards the overall outcome of a case, the system simply predicted whether the outcome of the lower court judgment would be affirmed or reversed, without giving reasons. The system appears not to have done so by any analysis of the *merits* of the matter, but through an analysis of the type of dispute, the court of origin, whether there was disagreement in the lower courts, whether oral argument was heard, and the tendencies of individual justices to reverse and/or dissent from decisions, as well as whether the type of issue had a "left/right" political slant.

In light of this, although the reported 70.2% accuracy is impressive, the limitations of the system underscore the difficulties in creating an AI system that can be used as a dispute resolution tool. The system does not appear to undertake legal analysis, but instead relies on knowledge of both the history of the matter in the lower courts, and the tendencies of the justice in question.

35 Smartsettle ONE at www.smartsettle.com/smartsettle-one (accessed 1 May 2021).

36 J. Hyde, "Mediator Claims Online Dispute First to Be Settled by Algorithm" (*The Law Society Gazette*, 25 February 2019) at www.lawgazette.co.uk/news/mediator-claims-online-dispute-first-to-be-settled-by-algorithm-/5069393.article (accessed 1 May 2021).

37 D. M. Katz, M. J. Bommarito II, and J. Blackman, "A General Approach for Predicting the Behavior of the Supreme Court of the United States" [2017] *PLoS one*, *12*(4) at https://arxiv.org/pdf/1612.03473.pdf (accessed 1 May 2021).

The system would therefore be of little use in predicting a judgment at first instance, where there will be no prior decision to rely upon, and (probably) a more limited case history showing the tendency of the relevant judge. Furthermore, the overall case outcome prediction was binary – "Affirm" or "Reverse".[38] This would be unlikely to be suitable for the prediction of judgments at first instance, which would be more likely to produce a broader range of potential outcomes both on individual issues and on quantum than a simple "win" or "lose". Further, even on its own terms, a 70.2% accuracy rating means the system will get a significant proportion of cases wrong. It is unclear how the parties could anticipate what features of a case may make it more or less amenable to accurate pre-determination.

While a broad-brush approach such as that taken by this system will doubtless be a tool for helping parties evaluate the strength of their case, we consider this is as far as the technology can go. For the foreseeable future, such systems will assist human lawyers in their evaluations, and not replace them, or human judges.

7.5 Conclusions

We have no doubt that the increased use of AI will dramatically impact maritime and trade law. The commercial applications of the technology will lead to a reappraisal of how existing case law applies to situations in which there has been no immediate human intervention. Alongside this, the lack of human beings to interrogate will likely lead to dramatic changes in how evidence is collected and interpreted.

Meanwhile in the practice of law itself, there will be new opportunities in assisting with the design of systems that will allow legal services to be more efficiently consumed – but at the price of a reduction in the flow of work that can be standardised and commoditised. Further, lawyers will not be the sole producers of such systems, but will need to work closely with those with the relevant technical skills.

Despite the above, however, we would not wish to overplay these developments. We have no doubt that the common law and rules of evidence can adapt to AI as they have adapted to all other technical advances since the industrial revolution. Further, we do not see the "human" skills involved in the practice of law, particularly those relating to how to make complex factual judgments, being replaced with mechanised versions any time soon.

However, doubt as to the capabilities of technology should be expressed with humility, and we must accept that the speed of technological change may well prove us wrong. On 1 February 2008, the University of Cambridge produced a press release on attempts of its researchers at the Cavendish Laboratory to statistically model the moves human players might make in the game of Go. At the time, the article noted that "there is currently no Go program that can play better than a beginner on the full-size board", and further that:

> some researchers . . . argue that a strong Go-playing computer is still decades away. However, the fact that humans learn the game with relative ease provides tantalising evidence that, one

38 Although it should be noted that there was one further option for predicting the votes of individual justices – "Other".

> day, it will be possible to create a strong Go program and, in so doing, perhaps increase our understanding of human intelligence.[39]

Less than one decade later, in May 2017, the AlphaGo program was not only "strong", but good enough to defeat Ke Jie, the world number one professional player in a three-game series.[40] Technology can sometimes move faster than even experts in the field consider possible.

39 "Modelling uncertainty in the game of Go" (*University of Cambridge Research News*, 01 February 2008) at www.cam.ac.uk/research/news/modelling-uncertainty-in-the-game-of-go (accessed 1 May 2021).

40 "AlphaGo China" (*DeepMind*) at https://deepmind.com/alphago-china (accessed 1 May 2021).

CHAPTER 8

Maritime intellectual property

Shining a light on the protection of disruptive technologies within the shipping industry

*Andrew Beale**

8.1 Introduction

JMW Turner's 1838 painting *The Fighting Temeraire* was an inspired choice of subject-matter for what was to become Britain's favourite painting.[1] Painted for exhibition at the National Gallery in London in the nearly completed Trafalgar Square, the *Temeraire* was famous as the ship which had come to the aid of Nelson's flagship *Victory* at the Battle of Trafalgar (1805). But the painting did more than tap into nostalgic sentiment. Employing artistic licence, Turner used the picture, the full title of which reads *The fighting Temeraire, tugged to her last berth to be broken up*, to convey the transition from the old age of sail to the new, and innovative, age of steam.

Steam may well have been a 'disruptive technology' in 1838: but the term itself was not to be coined until 1995. In their Harvard Business Review article '*Disruptive Technologies: Catching the Wave*',[2] Bower and Christiansen applied this name to new technologies which completely changed old ways of doing things, thereby disrupting, if not overturning, traditional established business models. Whereas the vast majority of inventions were only incremental improvements on existing technologies, the aspiration still remained to invent new technologies which were a radical improvement on old methods. Christensen went on to further work in this field, claiming that few technologies were intrinsically disruptive, and that it was generally the business models that the new technologies enabled which created the disruptive impact. In recognition of this he suggested 'disruptive innovation' as a more apt title, for example for the autonomous vessels of today.[3]

Gary Hamel famously said, "[S]omewhere out there is a bullet with your company's name on it. Somewhere out there is a competitor, unborn and unknown, that will render your strategy obsolete."[4] This quote is a sobering reminder that all shipping companies

* **OBE, Director of IP Wales and Member of the International Institute of Shipping and Trade Law (IISTL) Member of the International Institute of Shipping and Trade Law (IISTL)**

1 See 'The Greatest Painting in Britain Vote' (BBC *Radio 4 Today*, 2005) at www.bbc.co.uk/radio4/today/vote/greatestpainting/winner.shtml (accessed 21 August 2020).

2 J L Bower and C M Christensen. "Disruptive Technologies: Catching the Wave", Harvard Business Review 73(43) (1995).

3 Hellenic Shipping News worldwide, "Adopting Disruptive Technologies in the Maritime Industry" (*International Shipping News*, 22 November 2017) at www.hellenicshippingnews.com/adopting-disruptive-technologies-in-maritime-industry/ (accessed 21 August 2020).

4 B. Gaille, "65 Breathtaking Gary Hamel Quotes" (*Brandon Gaille Small Business & Marketing Advice*, 13 November 2016) at www.brandongaille.com/65-breathtaking-gary-hamel-quotes/ (accessed 16 September 2020).

DOI: 10.4324/9781003155195-8

need to engage with the innovation process if they are to survive beyond a single product/service life cycle.

The area of law used to protect innovation, disruptive or otherwise, is intellectual property (IP) law. This provides legal protection for differentiation achieved via branding and endorsements; new inventions or creations; and design or appearance. A patent may be the normal legal instrument of choice for protecting innovations within the shipping industry, but an IP strategy needs to go much further, since it should embrace the full range of intellectual assets of the business.[5] By way of illustration, the unique position occupied by shipping companies often means that they hold valuable data well beyond the kind of technical, scientific and commercial information you might expect. After all, as was observed eight years ago:

> if you really want an insight into how trends are shifting on trade flows – especially in Southeast Asia – it pays to listen to a shipping line. Maersk, the Danish operator which is the world's largest shipping line in terms of container traffic, sees what is carried on its ships daily, thanks to what is listed on bills of lading as cargoes loaded onto its ships.[6]

One of the major ways in which shipping companies can protect such valuable data is through the law of trade secrets. Whether they realise it or not, all shipping companies, whatever their size, are able to benefit from trade secrets; and there is indeed good evidence to suggest that they are increasingly coming to recognise the potential value of those secrets.[7]

The UK Trade Secrets Regulations 2018,[8] or TSR, brought into force the EU Trade Secrets Directive.[9] For the first time they enshrine in UK law a legislative definition for a trade secret: namely, any data which:

> (a) is *secret* in the sense that it is not, as a body or in the precise configuration and assembly of its components, generally known among, or readily accessible to, persons within the circles that normally deal with the kind of information in question,
> (b) has *commercial value* because it is secret, and
> (c) has been subject to *reasonable steps under the circumstances*, by the person lawfully in control of the information, to keep it secret.[10]

This definition offers what is generally a far broader understanding of what can constitute a trade secret than that previously applicable at common law, albeit the TSR recognises that in some cases the law of confidentiality may still be to offer "wider protection" in

5 "[D]ifferent forms of [IP] protection should be understood as an asset portfolio – you might want to diversify and spread your risk. A friend of mine who is the Vice President of a highly innovative company in the shipping industry appropriately describes intellectual property protection as a volume business: "You never know what will be the diamond in the rough" (Orly Lobel, "Filing for a Patent Versus Keeping Your Invention a Trade Secret" Harvard Business Review, 21 November 2013).

6 J. Grant, "Southeast Asia: Shipping Reveals Trade Secrets" *Financial Times* (London 11 March 2013).

7 "Some shipping companies remain reluctant to divulge their performance data about fuel origins, vessel operations, and the like because they consider it be their trade secrets" (B. Staton and H. Dempsey, "IMO Delegates Consider Shipping Industry Response to Calls for Cuts in Emissions" *Financial Times*, London, 10 November 2019).

8 Trade Secrets (Enforcement, etc.) Regulations 2018, SI 2018/597. These remain in force despite Brexit.

9 Formally, EU Directive 2016/943 on the protection of undisclosed know-how and business information (trade secrets) against their unlawful acquisition, use and disclosure.

10 Para. 2 (emphasis added).

the form of measures, procedures and remedies. Such wider protection is specifically preserved.[11]

The decision in *Trailfinders Ltd v Travel Counsellors Ltd*[12] presented the first opportunity for judicial scrutiny of the TSR. The approach adopted by HHJ Hacon in the Intellectual Property Enterprise Court was expressed thus:

> the substantive principles governing the protection of confidential information under English law, including that afforded by terms implied into contracts of employment and by equitable obligations of confidence, are unaffected by the Directive. However, the Directive shines an occasional light on those principles.[13]

In particular, he continued later, "the best guide to the distinction between information which is confidential and that which is not is now to be found in the definition of 'trade secret' in Article 2(1) of the Directive 2016/943".[14] This would imply a considerable updating of the established three-stage common law test for confidentiality: namely, that the information must have the necessary quality of confidence, have been imparted in circumstances importing an obligation of confidence (either expressly, or implicitly), and must have been used without authorisation to the detriment of the rights holder.

The author has previously highlighted that this marriage, which some might term a "shotgun wedding", between the new statutory definition of a "trade secret" and the old common law principles of confidentiality is not without its difficulties, in particular with HHJ Hacon's legal treatment of the terms "secret" and "reasonable steps".[15] To these we now turn.

8.2 The meaning of "secret"

The Trade Secrets Directive makes clear that the definition

> excludes trivial information and the experience and skills gained by employees in the normal course of their employment, and also excludes information which is generally known among, or is readily accessible to, persons within the circles that normally deal with the kind of information in question.[16]

This point came up in the *Trailfinders* case, where it was alleged that two employees of Trailfinders had wrongfully abstracted names, contact details and other information about Trailfinders clients, which they had used when moving to a rival business.

---

11 See Para. 3(1) of the same Regulations. This was recently described by Arnold LJ in *Shenzhen Senior Technology Material Co Ltd v Celgard LLC* [2020] EWCA Civ 1293; [2021] F.S.R. 1 at [29] as a "curious provision . . . it appears to be primarily intended to ensure that, if and in so far as English law prior to the implementation of the Trade Secrets Directive was more favourable to the trade secret holder . . . then that greater level of protection shall continue to be available." See A Beale, "Understanding TSR and its 'Curious Provision'" (*Institute of International Shipping and Trade Law*, 19 October 2020) at https://iistl.blog/2020/10/19/understanding-tsr-and-its-curious-provision/ (accessed 18 November 2020).

12 [2020] EWHC 591 (IPEC); [2020] I.R.L.R. 448. An appeal was dismissed by the CA at [2021] EWCA Civ 38.

13 See at [9].

14 See at [29].

15 A. Beale, "Inauspicious Start for the UK Trade Secrets Regulations at IPEC" (*Institute of International Shipping and Trade Law*, 27 May 2020) at https://iistl.blog/2020/05/27/inauspicious-start-for-the-uk-trade-secrets-regulations-at-ipec/ (accessed 21 August 2020).

16 Art. 14.

It was argued by the defendants that Trailfinders' client information had already been in the public domain and had therefore been "readily accessible". Trailfinders had held this client information on two systems: Viewtrail, an online portal used to record booking details, and Superfacts, a software system which recorded information about clients. The employees admitted using Superfacts to assemble a "contact book" about clients before they left, and accessing Viewtrail after they had left. But the judge took the view that the Trailfinder information had met the statutory threshold for being "secret". He also went further, however, citing Lewison LJ's observation in *Force India Formula One Team Ltd v Aerolab Srl*[17] that "[i]t is certainly not a defence [to an allegation of breach of confidence] that the person in breach of confidence could have obtained the information elsewhere if he did not in fact do so".[18]

8.3 The meaning of "reasonable steps"

In *Trailfinders*, HHJ Hacon felt able to find that although the protection might "not have been as rigorous as it should have been", nevertheless Trailfinders had clearly taken steps

> to ensure that the Client Information was not openly available to anyone by requiring the use of a password or, in the case of Viewtrail, limiting access to information to clients only if their name and booking reference was known.[19]

An important point is that he categorised the confidential information in question as information acquired during the normal course of employment (rather than some specific "trade secret").[20] This raises an important question: should more be demanded in the way of "reasonable steps" in the case of the latter than the former? In answering this question, and more generally dealing with the issue of what "reasonable steps" are expected, we may have to go further afield than Europe and the UK. The EU Directive and the UK Regulations based on it fairly directly mirror Article 39.2 of TRIPS,[21] an instrument which itself took inspiration from the American Uniform Trade Secrets Act,[22] a piece of legislation widely adopted by individual states. The USA for its part was active in promoting the implementation of the EU Trade Secrets Directive, specifically to protect US companies trading within the EU.[23]

In our IP Wales Paper entitled "The importance of keeping your company's trade secrets, secret"[24] we used such analysis to identify basic steps a board of directors generally might

17 [2013] EWCA Civ 780; [2013] RPC 36 at [35]. Briggs LJ and Sir Stanley Burnton agreed.

18 See at [72].

19 See at [73].

20 Because the latter under English jurisprudence enjoys a higher degree of confidentiality: see generally Goulding J's classification in *Faccenda Chicken Ltd v Fowler* [1985] 1 All ER 724. The Court of Appeal at [1987] Ch. 117 ultimately differed with Goulding J's analysis of where to draw the line between the two classes of information, but this does not affect the point in the text.

21 Formally, the WTO Agreement on Trade-Related Aspects of Intellectual Property Rights.

22 The text of the 1985 Uniform Act, which updates an earlier draft in 1979, can be viewed at www.wipo.int/edocs/lexdocs/laws/en/us/us034en.pdf (accessed 21 August 2020).

23 Department of Justice Report to Congress Pursuant to the Defend Trade Secrets Act (2018), 20. It is not surprising perhaps that the latest case to utilise the UK TSR should be an American company successfully seeking interim injunctive relief for the alleged misappropriation of its trade secrets by a Chinese competitor: see *Celgard LLC v Shenzhen Senior Technology Material Co Ltd* [2020] EWHC 2072 (Ch).

24 A. Beale and J. Mcfarlane, "The Importance of keeping your Company's Trade Secrets, Secret" (2020) IP Wales Articles at www.ipcybersecurity.com/free-guide-1 (accessed 21 August 2020).

undertake to seek to meet this "reasonable steps" threshold. These are as relevant to shipping as to other companies, and are described below.

8.3.1 Identification, labelling and steps taken to preserve trade secrets

US courts demand a noticeable degree of self-help from claimants in trade secrets cases. Protection is problematic if the claimant has done "nothing to differentiate its protective measures for the alleged proprietary trade secrets from those imposed on any other corporate information", rendering the alleged trade secret no "more confidential than all of the plaintiff's internal information".[25] Again, failure to label trade secrets data as such (or otherwise to designate it "confidential") has resulted in loss of trade secret status,[26] especially if there is already a company policy in place to do so[27] or where there is an analogous requirement in a non-disclosure agreement.[28] US cases indicate, moreover, that the reviewing of a data estate to identify and then label trade secrets is necessary but not sufficient: it is only a precursor to a requirement to take affirmative steps to preserve those secrets. Making employees,[29] and anyone else to whom the secret is revealed,[30] subject to a confidentiality agreement, is an obvious step: but it has been treated as merely a starting point, such that without more, the claim for trade secret protection is still likely to fail.[31] Measures that go beyond normal business practices may be required,[32] and not a mere reliance upon luck.[33] Poor enforcement of preservation measures may be worse than having no preservation measures at all. "As a matter of law, doing nothing to enforce a confidentiality agreement is not a reasonable effort . . . to maintain a trade secret."[34]

8.3.2 Staff education and training

An analysis of US trade secrets cases (at both federal and state level) has revealed that in over 90% of cases the defendant was either an employee, former employee or business partner.[35] One of the primary ways of revealing trade secrets to a competitor is, thus, employee misconduct concerning company information, whether intentional or negligent. To combat this and to satisfy the requirement of "reasonable steps", employers should, it

25 *Opus Fund Servs. (USA) LLC. v Theorem Fund Servs. LLC.*, No. 17-cv-923, 2018, WL 1156246 at 5 (N.D. Ill., Mar 5 2018).

26 *OTR Wheel Engineering Inc. v West Worldwide Services Inc.*, No. cv-14–085-LRS (E.D. Wash. Nov 30 2015).

27 *Call One v Anzine*, 2018 WL 2735089 (N.D. IL, 2018) at 9.

28 *Convolve, Inc. v Compaq Computer Corp.*, 812 F.3d 1313 (Fed. Cir. 2016).

29 *Mintel International Group Ltd v Neergheen* No. 8-cv-3939, 2010 WL 145786 (N.D. Ill. Jan 12 2010).

30 *ISC-Bunker Ramo Corp. v Altech, Inc.*, 765 F. Supp.1310 (N.D. Ill. 1990).

31 *Opus Fund Servs. (USA) LLC. v Theorem Fund Servs. LLC.*, No. 17-cv-923, 2018, WL 1156246 at 5 (N.D. Ill. Mar 5 2018).

32 *Gillis Associated Indus Inc. v Cari-All Inc.*, 206, Ill. App. 3d 184, 192, 564 N.E. 2d, 881, 886, (1st Dist. 1990).

33 *CMBB LLC v Lockwood Mfg Inc.*, 628 F. Supp. 2d 881 (N.D. Ill. 2009).

34 *Compuware Corp. v Health Care Serv Corp.*, 203 F. Supp. 2d, 952 (N.D. Ill. 2002) at 958.

35 Almeling et al. (2010) and (2011), cited in the Report prepared for the EU Commission under contract number MARKT/2011/128D, *Study on Trade Secrets and Confidential Business Information in the Internal Market* (Ref.Ares(2016)98815–08/01/2016, April 2013) at 107.

has been held, where appropriate, limit employee access on a need-to-know basis,[36] restrict ex-employee access[37] and institute a comprehensive system of employee training.[38]

8.3.3 Cybersecurity risk management

The duty to take reasonable steps extends in US jurisprudence to cybersecurity and risk management. Thus, US courts have at one time or another required a showing of proper identity and access management (for example, password protection, need-to-know access and secure server storage); of specific data security measures (for instance, USB use restrictions and distribution controls); of perimeter and network cyberdefences (such as firewalls, data encryption and online use restrictions); communication (pop-up warnings); and email monitoring.[39]

Whilst these steps obviously need only be "reasonable" and not "perfect",[40] an analysis of US cases indicates that even if an "impenetrable fortress" is not necessary, the burden placed upon the party alleging a trade secret can still be surprisingly onerous.[41] True, it has been said that "reasonable steps for a two-or-three person operated shop may be different from reasonable steps for a larger company";[42] and smaller businesses are not necessarily expected to undertake the same expensive steps to protect secrecy as larger ones.[43] Nevertheless, all businesses are required to have undertaken at least some affirmative action;[44] further, "a sophisticated party familiar with confidentiality agreements" can still act unreasonably when failing "to take any steps to maintain secrecy", irrespective of their small size.[45]

8.3.4 Generally

In the conclusion to our chapter we note that whilst the motive behind the EU Trade Secrets Directive might have been to level the playing field between European businesses and their counterparts in the USA and Japan,[46] there are differences. Notably, unlike the US Defend Trade Secrets Act 2016[47] the EU regime offers no criminal sanction against the misappropriation of trade secrets data. Indeed, UK criminal law has traditionally neither penalised the unauthorised taking of information as such nor considered data to be "property" within

36 *Stampede Tool Warehouse Inc. v May*, 272, Ill. App. 3d, 580, 209, Ill. Dec. 281, 651, N.E. 2d. 209 (1995).

37 *Orthofix v Hunter*, 55 F. Supp. 3d 1005, 1013 (N.D. Ohio 2014), 630 F. Appx 566 (6th Cir. 2015).

38 *Liebert Corp. v Mazur*, 357, Ill. App. 3d, 265, 293, Ill. Dec 28, 827 N.E. 2d, 909 (2005).

39 Center for Responsible Enterprise and Trade, *The Importance of Cybersecurity for Trade Secret Protection: Developments in trade secrets cases and the growing role of the NIST Framework* (2016).

40 *Learning Curve Toys v Playwood Toys Inc.*, 342 F. 3d, 714, 725 (7th Cir. 2003).

41 *Boston Scientific Corp. v Dongchul Lee*, No. 1:13-cv-13156-DJC (D. Mass.).

42 *Puroon Inc. v Midwest Photographic Res Ctr Inc.*, No. 16-cv-7811, 2018, WL 5776334 (N.D. Ill. Nov 2 2018) at 7.

43 *Elmer Miller Inc. v Landis*, 253 Ill. App. 3d 129, 134, 625 N.E. 2d, 338, 342 (1st Dist. 1993).

44 *Jackson v Hammer*, 274, Ill. App. 3d, 59, 67, 652 N.E. 2d, 809, 815 (4th Dist. 1995).

45 *Fail-Safe LLC v A.O. Smith Corp.*, 674, F. 3d, 889 (7th Cir. 2012).

46 See Commission Response to the Report prepared for the EU Commission contract number MARKT/2011/128D, *Study on Trade Secrets and Confidential Business Information in the Internal Market* (Ref. Ares(2016)98815–08/01/2016, April 2013).

47 130 Stat. 376.

the meaning of the Theft Act 1968.[48] So the criticism of Lord Boyle that "it is not too much to say that we live in a country where . . . the theft of the board room table is punished far more severely than the theft of the board room secrets" holds as true today as it did more than fifty years ago.[49]

8.4 Maritime trade secrets: the way forward

The maritime field is awash with examples of major cyberattacks.[50] We have seen that a shipping company cannot treat the misappropriation of its trade secrets as "theft" under UK criminal law, but can it at least engage in more offensive forms of cyber defence of their trade secrets, measures that go beyond mere passive defensive actions, such as deploying anti-virus software?

The author has previously noted that retaliatory hacking by a shipping company as the victim of a cyberattack is

> likely to be deemed illegal in the UK under the Computer Misuse Act (CMA) 1990, which makes it a criminal offence to gain unauthorised access to a third party's computer or data or otherwise impair the operation of a third party's computer . . . [U]nder pressure from Civil Rights Groups, the UK Government has found it expedient to introduce new legislation to "clarify" that [only] GCHQ, Intelligence Officers and the Police possess the lawful authority to engage in hacking (pre-emptive or retaliatory) without criminal liability – thereby ensuring the "Offensive Cyber" capabilities set out in the UK National Cyber Security Strategy 2016–2021.[51]

In our recently published *IP Wales Guide to Cyber Defence* we record that

> since 2004, a UN Group of Governmental Experts (UN GGE) has sought to expedite international norms and regulations to create confidence and security-building measures between member states in cyberspace. In a first major breakthrough, the GGE in 2013 agreed that international law and the UN Charter is applicable to state activity in cyberspace. Two years later, a consensus report outlined four voluntary peace time norms for state conduct in cyberspace: states should not interfere with each other's critical infrastructure, should not target each other's emergency services, should assist other states in the forensics of cyberattacks, and states are responsible for operations originating from within their territory.[52]

However, with international tensions currently running high between the USA and China this process has become stalled. In giving evidence to Parliament on "Cybersecurity in the UK" Sir Mark Sedwill, as the (then) Cabinet Secretary, Head of the UK Civil Service and UK National Security Advisor revealed that the UK is looking for a "coalition of the willing" to take this cyberspace governance issue forward, whilst at the same time highlighting the parallel with the law of the sea two hundred years ago.[53]

---

48 See e.g. *Oxford v Moss* (1979) 68 Cr App R 183 Section 4(1) Theft Act 1968.

49 Hansard (HC) 13 December 1968, vol 775, col 806.

50 See e.g. COSCO (2018); Maersk (2017); AIS & Spoofing (2017); Ports – Antwerp (2011); Barcelona & San Diego (2018); Oil rigs – Mexico; Africa.

51 A. Beale, "A Right to Bear Cyber Arms?" *Cambridge International Law Journal*, 19 November 2019 at http://cilj.co.uk/2019/11/19/a-right-to-bear-cyber-arms/> (accessed 16 September 2020).

52 F. Manig and A. Beale, "Briefing Guide: Cyber Defence" 8 (2018) at www.ipcybersecurity.com/free-guide-1 (accessed 16 September 2020).

53 Public Accounts Committee Oral evidence of Sir Mark Sedwell: *Cyber Security in the UK*, HC 1745 [1st April 2019] Q93.

In drawing this parallel with the law of the sea, Sir Mark may have had in mind academic papers such as *Cybersecurity and the Age of Privateering: A Historical Analogy*, in which the author asserted:

> 1. Cyber actors are comparable to the actors of maritime warfare in the sixteenth and seventeenth centuries. 2. The militarisation of cyberspace resembles the situation in the sixteenth century, when states transitioned from a reliance on privateers to dependence on professional navies. 3. As with privateering, the use of non-state actors by states in cyberspace has produced unintended harmful consequences; the emergence of a regime against privateering provides potentially fruitful lessons for international cooperation and the management of these consequences.[54]

If a "coalition of the willing" is to be achieved, especially between the five-eyes community (intelligence co-operation programme between the United States, United Kingdom, Australia, Canada, New Zealand) to take this cyberspace governance issue forward, the author would suggest a common platform now needs to be agreed on the more aggressive defensive cyber actions private companies should and should not be permitted to conduct in the defence of their trade secrets.

54 F. Egloff, "Cybersecurity and the Age of Privateering: A Historical Analogy" (2015) University of Oxford Cyber Studies Programme, Working Paper Series No. 1 at www.politics.ox.ac.uk/materials/publications/14938/workingpaperno1egloff.pdf (accessed 16 September 2020).

CHAPTER 9

The human element in autonomous shipping

*Zoumpoulia Amaxilati**

9.1 Introduction

Speaking about the human element in autonomous shipping may seem paradoxical: how will human intervention be at all relevant once maritime autonomous surface ships (MASS) will enter into use? It is certainly true that when fully autonomous ships are introduced, certain types of human element will inevitably become obsolete. At this highest degree of autonomy, for example, traditional on-board personnel will be replaced by an operating system capable of making decisions and determining actions by itself without any human intervention[1] (though even here it is worth noting that the complete absence of any remote and/or on-board human control will not preclude the possibility of human observation in the form of on-board or shore-based security personnel).[2]

However, until this fully autonomous stage is reached, the human element will remain actively involved throughout the whole field of MASS. At the lowest degrees of autonomy, as described by the IMO, human intervention is expected not only in the production stage, where technical systems are to be designed, constructed and tested by human beings, but also in the operational stage, where the relevant vessels will remain controlled by seafarers on board and/or by remote operators. Autonomy degree one, for example, will introduce fairly minor differences between MASS and conventional ships, in a sense that seafarers are always going to be on board to operate and control automated shipboard systems and functions. By contrast, autonomy degree two will feature a novel combination of remote and on-board operational control, which will in turn be replaced by a completely remotely controlled operational system in autonomy degree three.

Furthermore, it is expected that, during the initial phase of autonomous shipping, MASS will have to interact with conventional ships operated by human beings.[3] Finally, human involvement is likely to be maintained in a wide range of ancillary operational matters, such as cargo loading and discharge operations, ship repair and supply, and inspection and certification procedures, to name but a few.[4]

* **Lecturer in Shipping and Trade Law, Institute of International Shipping and Trade Law, Swansea University**

1 G. Wright, *Unmanned and Autonomous Ships: An Overview of MASS* (1st edn, Routledge, 2020), 42.
2 Ibid.
3 S. Ahvenjarvi, "The Human Element and Autonomous Ships" TransNav (2016) 10: 517, 518.
4 Ibid.

DOI: 10.4324/9781003155195-9

9.2 The aim of this chapter

The various manifestations of the human element on MASS spark a wide range of legal issues from the point of view of regulation. These include issues of product liability for the builders and manufacturers of autonomous ships and their components, or the programmers of the software; compulsory pilotage issues; and issues of carrier liability, to name but a few.[5] It is beyond the scope of this chapter to provide a detailed discussion of all of these. Instead, its aim is to look at regulatory issues regarding MASS controlled by remote operators and to make proposals as to how these gaps in the regulatory framework might be filled.

The first challenge regarding remotely controlled MASS operations is to determine the legal position of remote operators. Can they be considered seafarers? If not, then what should their legal status be? Should they be regarded as employees of the shipowner or independent contractors? In either case, can they be considered as functional equivalents of seafarers? The legal position of remote operators will be relevant to questions relating, for instance, to their employment rights; to their civil and criminal liability in the event of maritime accidents; to their rights to limit liability for maritime claims; and their right to arrest the relevant vessel in aid of claims for any unpaid wages. These questions will be the focus of the first part of this chapter.

The IMO's strategic plan for 2018–2023 sets a clear aim to "integrate new and advancing technologies in the regulatory framework", which they say requires balancing "the benefits derived from such technologies against safety and security concerns, the impact on the environment and on international trade facilitation, the potential costs to the industry, and finally their impact on personnel, both on board and ashore".[6] Effectively, this means that, in order to introduce MASS operationally, standards with regard to safety, security and protection of the environment should be at least comparable with those provided by the relevant instruments for conventional ships. Furthermore, any personnel involved in MASS operation, whether remote or on board, should be appropriately qualified and experienced to operate MASS safely. Unlike seafarers, for the qualification of which there is a long maritime tradition and clear requirements are established under the International Convention on Standards of Training, Certification and Watchkeeping for Seafarers 1978 (the STCW), remote operators will be a new type of professional, with new requirements in terms of skills. As a result, the potential challenges regarding the training and certification standards for remote operators will be considered in the second part of this chapter.

Finally, a wide use of MASS controlled by remote operators will have the potential to challenge traditional approaches to maritime liability. Remotely controlled ship operations will introduce a new class of individuals who will be expected to play the most important role in the navigation of those vessels. The presence of remote operators, in consequence,

5 See, for example, F. Stevens, "Carrier Liability for Unmanned Ships" in B. Soyer and A. Tettenborn (eds), *New Technologies, Artificial Intelligence and Shipping Law in the 21st Century* (Informa Law from Routledge, 2020); A. Tettenborn, "Product Liability Goes High-Tech" in B. Soyer and A. Tettenborn (eds), *New Technologies, Artificial Intelligence and Shipping Law in the 21st Century* (Informa Law from Routledge, 2020); and L. Carey, "All Hands Off Deck? The Legal Barriers to Autonomous Ships" CML Working Paper Series, No. 17/06, August 2017 at https://law.nus.edu.sg/wp-content/uploads/2020/04/011_2017_Luci-Carey.pdf (accessed 18 January 2020).

6 IMO Resolution 1110 (30) at wwwcdn.imo.org/localresources/en/About/strategy/Documents/A%20 30-RES.1110.pdf (accessed 15 January 2021).

will raise questions such as these: how criminal liability will be apportioned between the master and remote operators when a collision takes place; whether the master or the remote operators (or both) will be liable in tort for damage or loss of property or for personal injury or loss of life caused as a result of a collision; whether the shipowner can be made liable, directly or vicariously, in these circumstances; and whether a degree of strict liability might be a viable solution to address the potential challenges. This area will be the last issue to be discussed, at the end of the chapter.

9.3 The legal status of MASS remote operators

The first question to consider is whether it will be possible for MASS remote operators, by the very nature of their work, to be considered seafarers. In other words, is a seafarer necessarily an individual who renders service on board a ship? This is currently governed by the Maritime Labour Convention 2006 (the MLC), which prescribes a set of rights and principles governing the minimum requirements for seafarers to work on board a ship, the conditions of employment for seafarers, and the minimum standards for decent working and living conditions on board ships. The MLC is centred on the premise that a seafarer is any person who is employed or engaged or works in any capacity on board a ship to which it applies.[7] Thus, it seems, the absence of any need for on-board attendance could potentially negate the application of the MLC regime.

Similar definitions of the term "seafarer" are adopted in domestic shipping legislation. In the UK, for example, a seafarer is described as "every person (except masters and pilots) employed or engaged in any capacity on board any ship".[8] Furthermore, emphasis added, for an individual to be considered seafarer, his or her "normal place of work must be on board a ship".[9] Most certainly on-board presence is required by virtue of this definition, and those similar to it. Such definitions will arguably preclude MASS remote operators from obtaining the legal status of seafarer given that, while they will be charged with duties and responsibilities relating to the navigation of MASS, they are expected to be entirely shore-based.[10]

There may, of course, be arguments in favour of classifying MASS remote operators as seafarers. Presumably, the only way to achieve this would be to adopt a broader definition of a "seafarer" by reference to functions normally performed by seafarers in respect of the ship's navigation, unconstrained by any requirement of on-board attendance. However, it is arguable that such an approach would be undesirable, as it would disregard the reasons behind the recognition of seafarers as a special category of workers requiring special protection. This special categorisation and protection of seafarers has customarily been recognised by the IMO and the International Labour Organisation (the ILO). In its preamble, the MLC states that such protection is necessary given the global nature of the shipping industry. However, unlike traditional seafarers MASS remote operators will not

7 MLC, Art. 2.

8 Merchant Shipping Act 1995 (hereafter the MSA), s. 313.

9 Merchant Shipping (Maritime Labour Convention) (Minimum Requirements for Seafarers etc.) Regulations 2014 (SI 2014/1316), Reg. 2(1).

10 S. Baughen, "Who Is the Master Now? Regulatory and Contractual Challenges of Unmanned Vessels" in B. Soyer and A. Tettenborn (eds), *New Technologies, Artificial Intelligence and Shipping Law in the 21st Century* (Informa Law from Routledge, 2020) 138.

interact with different jurisdictions in the course of their employment, since they will be exclusively located ashore; if so, it can plausibly be argued that they do not need the same degree of protection.

However, even if this is incorrect and MASS remote operators do require special protection, it remains hard to see how this special protection should be the same as that accorded to seafarers. Individual rights granted to seafarers aim to serve the peculiarities of the seafaring profession. Take, for example, the right to repatriation. It ensures that those seafarers, who are unable or cannot be expected to carry out their duties under their employment agreements or whom employment agreements expire or are terminated abroad, are able to return home.[11] Clearly, this right will have no practical significance for MASS controlled by remote operators, where those operators will work from their home countries anyway.

Furthermore, in a number of respects seafarers are treated as favoured litigants in respect of civil claims. For example, they are guaranteed pre-judgment security in respect of claims for wages, since under Arts. 1 and 3 of the International Convention Relating to the Arrest of Sea-Going Ships 1952 (given effect in more than 70 countries, including the UK),[12] they can arrest either the ship in respect of which their claim arose, or any sister-ship, and may do so even though the ship arrested may be ready to sail.

This special treatment of seafarers is encountered at national level as well. In the UK, courts have long recognised the existence of a maritime lien for seafarers' wages[13] providing a claim on the ship for service done to the ship, independently of any contract or personal liability on the part of the owner; its effect is to give security for remuneration for services rendered in relation to, or in connection with, the relevant vessel.[14] Could it be argued that such a lien should equally be available to a MASS remote operator? There are convincing arguments both ways. Against such a lien is the point that traditionally on-board attendance has been necessary for the lien to arise. On the other hand, there seems to be no reason for any dogmatic requirement for service to have been rendered on board the ship, as long as it was provided in relation to, or in connection with, the ship. Indeed, in *The Ever Success*,[15] Clarke J said this:

> a master or a seafarer is entitled to wages and thus to a co-extensive maritime lien if he renders the service appropriate to his rank. . . . He must be part of the crew of the ship, but need not necessarily render the service on board the ship or live on board the ship, but the service must be in a real sense referable to the ship and the service must be rendered during a period when the particular claimant can fairly be said to be part of the crew of the ship.[16]

Therefore, the most significant question in this respect is whether MASS remote operators can be considered part of the ship's crew of those vessels. Simon Baughen argues that MASS remote operators cannot be said to be part of the ship's crew, as there will be

11 MLC, Reg. 2.5.

12 Note that a similar right is provided under Art. 1(o) of the International Convention on Arrest of Ships 1999, ratified by a few (though not very many) jurisdictions, including Albania, Algeria, Benin, Bulgaria, Congo, Ecuador, Estonia, Latvia, Liberia, Spain, Syrian Arab Republic and Turkey.

13 *The Bold Buccleugh* (1851) 7 Moo. PC 267.

14 *The Castlegate* [1893] AC 38.

15 [1999] 1 Lloyd's Rep. 824.

16 See [1999] 1 Lloyd's Rep 824, 832.

no crew.[17] Certainly, such an approach would preclude MASS remote operators from benefiting from a maritime lien for any unpaid wages when MASS will operate at autonomy degree three.[18] However, it is argued, it would leave a window open for the application of the maritime lien for any unpaid wages to be extended to MASS remote operators when a combination of on-board and ashore operational control is used.

Importantly, however, arguments against the extension of the maritime lien to MASS remote operators can be based on the fact that the underlying reasons justifying the maritime lien do not apply to them. In the context of remotely controlled ship operations, it is argued, it would be hard to establish the necessary close connection between services rendered by the operator and any particular ship, since typically multiple operators would be involved in the operation of particular ships, and conversely remote control centres would handle multiple vessels at the same time.[19] A potential way around this would be legislation specifically making a maritime lien available to MASS remote operators; but this would be problematic, and seems a far-fetched possibility.

Seafarers, finally, enjoy special protection when a ship sinks or is lost. This has been enshrined in the MLC, Regulation 2.6 of which states that "in every case of loss or foundering of any ship, the shipowner shall pay to each seafarer on board an indemnity against unemployment resulting from such loss or foundering".[20] Furthermore, seafarers are entitled to adequate compensation in the case of injury or death arising from the ship's loss or foundering.[21] Clearly, the latter will not be relevant to any personnel located ashore. However, it may be argued that, if a MASS is lost, a remote operator should be compensated for any resulting loss of employment. Understandably, this argument cannot succeed if the relevant obligation is particularised by the requirement of on-board attendance.[22] But, even if there were no such requirement, a right to compensation would still cause difficulties in cases where remote control centres were responsible for handling more than one ship in the specific circumstances.

It seems, therefore, that there are good arguments for not regarding MASS remote operators as seafarers. However, the obvious question which then presents itself is what their legal status should be. It is suggested that they should be treated as a novel category of maritime professionals. Regarding their legal status, it is submitted, there are two possibilities, which require further consideration.

One option would be for them to obtain the legal status of shore-based employees. In this way, their employment rights will be regulated by the laws of the country in which they ordinarily render their services. The risk here, however, is that MASS remote control centres may be installed in countries with relaxed employment laws, in order to reduce

17 S. Baughen, "Who Is the Master Now? Regulatory and Contractual Challenges of Unmanned Vessels" in B. Soyer and A. Tettenborn (eds), *New Technologies, Artificial Intelligence and Shipping Law in the 21st Century* (Informa Law by Routledge, 2020) 138, 139.

18 Autonomy degree three covers purely remotely controlled ship operations without any seafarers being available on board.

19 The possibility of remote control centres handling multiple MASS has been acknowledged in the Rolls-Royce project and the Maritime Unmanned Navigation through Intelligence Networks (MUNIN) project. For more details, see G. Wright, *Unmanned and Autonomous Ships: An Overview of MASS* (1st edn, Routledge, 2020) 95.

20 MLC, Standard A2.6.

21 MLC, Regulation 2.6.

22 Ibid.

operational costs. There are also fundamental questions as to who their employer should be. The shipowner is probably the most natural answer to this question. However, consider a MASS under a charter by demise. Traditionally, in such case, the master and crew are employed by the charterer.[23] Can this be equally true for remote operators controlling a MASS under a charter by demise? It is strongly arguable that shipowners may not easily trust bareboat charterers with the navigational control of their vessels, given that the value of MASS is expected to be significantly higher compared to the average value of a conventional ship. Further, the prospect of the shipowner being the employer may be problematic whenever remote control centres will be responsible for multiple MASS owned by different shipowners.

If so, it may be better to grant MASS remote operators the status of independent contractors. Furthermore, such a view may be preferred by owners of MASS for economic reasons. However, as will be explained further on, this option can equally be problematic in the field of maritime liability.[24]

9.4 Training and certification standards for MASS remote operators

Drawing on the conclusion that MASS remote operators cannot be considered seafarers, even though they perform parts of the duties traditionally carried out by seafarers, the next issue to consider is that of the relevant training and certification standards. Hitherto there has been no international legal instrument regulating the training and certification standards for MASS remote operators.[25] Furthermore, the STCW Convention, which prescribes minimum qualification and training standards for seafarers working on board seagoing ships,[26] does not seem to offer any guidance in this area. This is because it covers conventional ship operations and does not purport to govern any other modes of operation.[27] However, this does not mean that it will have no bearing whatsoever to MASS controlled by remote operators. In such cases, the STCW Convention will continue to be relevant to the extent that seafarers will remain available on board those vessels.

As explained above, the requirements for, and capabilities needed of, remote operators will integrate existing maritime skills with new skills to cover the aspects of remote and autonomous functions relating to the operation of MASS from remote control centres.[28] This is necessary since remote operators will be responsible for all aspects of safe navigation.[29] Operators should therefore be familiar with the technology involved, particularly any unique features and equipment present on MASS, and be able to take best advantage

23 M. Davis, *Bareboat Charters* (2nd edn, Informa Law by Routledge, 2005) Chapter 1, Section 1.1.

24 This issue will be considered in the third part of this article.

25 However, training is an area that has been addressed by the industry. See, for example, the Maritime Autonomous Surface Ships (MASS) UK Industry Conduct Principles and Code of Practice which is produced by the Maritime UK Autonomous Systems Regulatory Working Group (MASRWG). While it is not a legal text, this Code of Practice seeks to provide practical guidance for the design, construction and safe operation of autonomous and semi-autonomous MASS less than 24m in length while the more detailed regulatory framework for MASS is developed under the MSA 1995. The third version of this Code of Practice was published in November 2019.

26 STCW, Art. III.

27 R. Veal and M. Tsimplis, "The Integration of Unmanned Ships into the *lex Maritima*" [2017] LMCLQ 303, 323 to 323.

28 G. Wright, *Unmanned and Autonomous Ships: An Overview of MASS* (1st edn, Routledge 2020) 91.

29 Ibid., 96.

of automated shipboard systems and functions while simultaneously maintaining knowledge and experience equivalent to that of seafarers.[30] Thus in principle, the potential training and certification standards must cover aspects of conventional seafarer training as well as education and training in communications and software technology advances and the specialised protocols surrounding the operations of MASS.[31]

Further lessons regarding the substance of the potential training and certification standards for remote operators can be learned from the existing regime governing the standards of training and certification for vessel traffic service (VTS) personnel.[32] Similar to what will be expected from remote operators of MASS, VTS personnel control maritime traffic, with a view to preventing dangerous situations, collisions, grounding of vessels and oil spill accidents, within the limits of a port.[33] In performing their duties, they rely on knowledge and skills as regards navigation and technical handling of VTS systems, such as communications technology, vessel traffic management information systems (VTMIS), radar, audio, video and other sensors, VHF and DF radio, tracking systems, including automatic radar plotting aids (ARPA) and automatic identification systems (AIS), and other new technologies, as appropriate.[34]

VTS operator training consists of a range of modules covering language skills, traffic management, equipment, nautical knowledge, communications co-ordination, VHF radio, personal attribute and emergency situations.[35] Prospective candidates are expected to meet the minimum entry requirement of the STCW II/1 Officer of the Watch (OOW) Deck or equivalent;[36] if they do not, then while they may still attend the course, they have to undertake any additional radar, ARPA, nautical knowledge and VHF radio training which may be required.[37] Owing to the unique features and equipment of MASS technology, it would not be practically possible to use any available model courses and guidelines relating to the training of VTS personnel as such. However, it is submitted, the principles and general directions could form the groundwork for regulators to build standards of training and certification for remote operators to meet their specialised needs.

In terms of procedure, the most appropriate way of addressing the potential challenges regarding the training and certification standards for MASS remote operators would be to adopt a new legal instrument in the form of a convention. This could be negotiated and drafted through the regulatory process of the IMO, and its structure could follow that of other similar instruments, particularly the structure of the STCW Convention.[38] The new

30 Ibid.

31 R. Veal & M. Tsimplis, "The Integration of Unmanned Ships into the *lex Maritima*" [2017] LMCLQ 303, 323 to 323.

32 IMO Resolution A 857 (20).

33 IMO Resolution A 857 (20) Annex 1, Guidelines 1 and 2.

34 Ibid., Annex 2, Figure 3.

35 International Association of Marine Aids to Navigation and Lighthouse Authorities (IALA) Guidelines G1156 (edn 1.0, December 2020) Guideline 6. For the UK, in particular, see Marine Guidance Note (MGN) 434 (M+F) (Amendment 1) Section 8.

36 IALA Guidelines G1156 (edn 1.0, December 2020) Guideline 6.1. For the UK, in particular, see Marine Guidance Note (MGN) 434 (M+F) (Amendment 1) Section 4.

37 Ibid.

38 The STCW Convention consists of the Articles, the Annex and the Code. The Convention Articles, the Annex and Part A of the Code are mandatory. Part B of the Code is recommendary in nature and provides guidance to assist those involved in educating, training or assessing the competence of seafarers or who are otherwise involved in applying the provisions of the STCW Convention.

convention would thus be the equivalent of the STCW Convention for remotely controlled MASS operations.[39] The former would supplement the latter, and both would operate side-by-side whenever a hybrid model of remote and on-board operational control was used. In other words, the convention on standards of training and certification for MASS remote operators would be relevant to those individuals located exclusively ashore and charged with operational duties and responsibilities whereas the STCW Convention would continue to apply when seafarers would be on board and available to take control of and operate shipboard systems and functions.

On a final note, it seems appropriate to highlight that, in every case of this short, it is not enough to merely prescribe requirements for training and qualification. It is also necessary to set out clear standards, preferably at an international level, for the authorisation and accreditation of the institutions that will provide the said training and qualification. This is particularly important in cases like the present where there is no previous experience, if the main concern is to ensure that MASS controlled by remote operators will function at least as safely as conventional ships. Having international standards for training providers, it is argued, will preclude the emergence of training organisations of convenience, which will jeopardise the safety standards for remotely controlled MASS operations.

9.5 The liability of MASS remote operators in respect of collisions

Turning now to issues surrounding the liability, both criminal and civil, of MASS remote operators when a collision takes place, it may be useful to consider the following hypothetical example. A conventional ship exits a port in a coastal state. A few miles outside, sailing in breach of Rule 16 of COLREGS, she collides with a vessel crossing on her starboard side, damaging the latter and injuring two of her crew. According to traditional approaches to maritime liability, in such circumstances the shipowner and the master of the "guilty" ship would bear responsibility and be subject to any resulting liabilities.

In the UK, the master can be guilty of an offence in respect of the negligent operation of the ship.[40] In practice, of course, the master is not always the individual who is in charge of the navigation of the vessel. However, the master, holding an overriding authority on board the ship, is the person responsible for decisions on issues relating to safety and security on board and the protection of the marine environment.[41] In addition, any other person for the time being responsible for the conduct of the ship, together with the shipowner, can face criminal charges for the same offence if a ship which is within the UK territorial waters, or is a UK ship wherever she may be, is negligently operated.[42]

The master can also be liable in tort for any physical damage caused as a result of the negligent navigation of the ship.[43] In order for this claim to succeed it must be established

39 The idea of creating a sister Convention to the STCW, which would cater for the particular needs of MASS controlled by remote operators, has also been suggested by Luci Carey. For more details, see L. Carey, "All Hands Off Deck? The Legal Barriers to Autonomous Ships" CML Working Paper Series, No. 17/06, August 2017 at https://law.nus.edu.sg/wp-content/uploads/2020/04/011_2017_Luci-Carey.pdf (accessed 18 January 2020).

40 Merchant Shipping (Distress Signals and Prevention of Collisions) Regulations 1996 (SI 1996/75), Reg. 6.

41 International Convention for the Safety of Life at Sea 1974 (SOLAS), Chapter V, Reg. 34–1.

42 Merchant Shipping (Distress Signals and Prevention of Collisions) Regulations 1996, Reg. 6.

43 It may be worth noting here that, in such circumstances, the master will most likely be entitled to limit his/her exposure to liability, as Article 1(4) of the International Convention on Limitation of Liability for Maritime Claims 1976 (LLMC) provides that: 'If any claims set out in Article 2 are made against any person for whose act,

that the defendant owed the claimant a duty of care; that the duty was broken by failing to meet the standard of reasonable care; that the breach caused the damage; and that the damage was not too remote. At common law, it is well settled that the master owes a duty of care to shipowners, crew and passengers of other vessels, who may foreseeably suffer physical damage by the negligent conduct of a vessel,[44] and that the master must show the standard of care of the hypothetical reasonable master,[45] with failure to comply with international navigation rules very strong evidence of negligence.[46]

In practice, of course, a claim of compensation against the master personally would be of limited practical value, as the master probably lacks the financial means to satisfy a judgment. A claimant is much more interested in bringing the claim against the shipowner, who would be better able to satisfy a judgment and more likely to be insured. There are two ways to achieve this. One is to bring a claim under the doctrine of vicarious liability against the owner, who will be liable for the master's negligence.[47] The other is to pursue a claim against the shipowner for systemic negligence or corporate failure; such a claim will succeed if it can be established that the shipowner's systems or procedures fell below the standard of a hypothetical reasonable shipowner. While the former can be taken for granted in collision situations, the same cannot be said about the latter, as it may be difficult to show serious flaws on the shipowner's organisation or system of work.

Let us assume now that our example involved a MASS controlled by remote operators and not a conventional ship. How would the liability matrix be formulated? It is argued that, inevitably, the involvement of remote operators in the navigation of those ships would have the potential to challenge the effectiveness of the current regime. Where, for example, this hybrid model of on-board and remote operational control is used, there may be questions as to whether the master should incur liabilities, both criminal and civil, for the ship's negligent conduct should lie with the master rather than the remote operators, if the latter control the shipboard systems and functions, and make the navigational decisions of the ship.

In the context of criminal liability, it is submitted, there are three possible answers to such questions. One option would be to accept that the master should retain an overriding authority, being the person responsible for decisions in relation to safety and security on board the ship and the protection of the marine environment.[48] Thus, any resulting criminal liability would remain on the master, together with the shipowner.[49] On such an approach, MASS remote operators would not face criminal liability, unless there was evidence that at the time of the collision they were in charge of the vessel, in which case they would be criminally liable.[50] Clearly, this view would be preferred by coastal states, as the

neglect or default the shipowner or salvor is responsible, such person shall be entitled to avail himself/herself of the limitation of liability provided for in this Convention'.

44 *The Hua Lien* [1991] 1 Lloyd's Rep 309, 328–332.

45 *The Dundee* (1823) 1 Hagg 109, 120.

46 *The Nowy Sacz* [1979] QB 236.

47 This is because the master would be the shipowner's employee, whose negligence was committed during the course of his/her employment, or an individual who shares with the shipowner a relationship which is "akin to employment", and whose negligence was connected to that relationship.

48 SOLAS, Chapter V, Reg. 34–1.

49 However, the enforcement of such criminal charges against the shipowner might be difficult if the latter were based abroad.

50 Of course, it would not be possible for coastal States to enforce such charges if the remote control centre is located abroad.

enforcement of any criminal charges against the master would remain possible when there was a collision within their territorial waters. However, it is suggested that this is unfair to the master, in that it would subject him or her to criminal prosecution arising out of functions and tasks unique to MASS, in relation to which he or she might have limited knowledge or experience.

An alternative would be to allocate responsibility between the master and remote operators by reference to specific functions and tasks. Generally speaking, the master would be responsible if manual operation was in place, and remote operators in so far as they were controlling the shipboard systems and functions. Thus in a collision any resulting criminal liability would be apportioned between the master and remote operators accordingly. Although this arrangement would offer a more balanced solution, it could still be problematic. It might not always be easy to draw a line between remote and manual operation control; further, even if the line could be drawn there could be resulting difficulties and delays in investigating maritime accidents. In addition, it would be difficult for coastal states to enforce criminal charges against operators abroad: a factor which, as will be suggested below, could well cause such states to adopt specific conditions for the entry of MASS controlled by remote operators into their ports.

The last option would be a regime of concurrent criminal liability covering the master, the remote operator, and the shipowner. However, as with the arrangements just discussed, this view could be problematic for many of the same reasons. Put in simple terms, it could well lead to scapegoating of seafarers on the spot who would face possibly unfair criminal pursuit while the remote operators, outside the jurisdiction, would be effectively immune. Theoretically this could be counterbalanced by recognising that it should be a defence for the master to show that, for the relevant time, they were not responsible for the conduct of the ship: but the practical difficulties attaching to such a defence should not be underestimated.

In so far as civil liability is concerned, it is suggested that the most suitable answer to such queries would be to recognise a concurrent duty of care on both the master and remote operators to exercise reasonable skill and care for the safe navigation of the vessel: where there was overlapping responsibility, the eventual delineation of liability between the master and remote operators would be a question of fact to be determined by the courts. Hence a claimant would be able to bring concurrent claims in negligence against the master and remote operators. Of course, as seen earlier, such claim would be accompanied by a claim against the shipowner for vicarious liability.[51]

On the other hand, where there are no seafarers on board and the ship is controlled entirely remotely, different questions may arise. For criminal liability purposes, in particular, the procedural difficulties in respect of the enforcement of any criminal charges against remote operators would be severe. This, taken in conjunction with the vindication and security purposes achieved by those criminal charges, could lead coastal states to set conditions for the entry of MASS controlled by remote operators into their ports.[52] Such conditions could require of MASS to shift into a hybrid method of remote and on board operational control or a conventional method of manned operation before entering ports, so

51 The extent to which the shipowner might be found vicariously liable for the remote operators' negligent navigation of the ship will be discussed in the following paragraphs.

52 United Nations Convention on the Law of the Sea 1982 (UNCLOS), Art. 25.

that coastal states would be able to identify an individual who would be criminally liable for the ship's actions. However, this should not be encouraged, as it would effectively mean that the scapegoating of seafarers would continue.

As regards civil liability, the most important question would be whether the shipowner could be regarded as vicariously liable for the remote operators' negligent navigation of the ship.[53] The answer to this question would very much depend on the legal status of the remote operators. If they were employed by the shipowner, then the latter would be vicariously liable for their fault. But if they were independent contractors, vicarious liability would be precluded, and the shipowner could only be found liable if systemic negligence were established: for instance, by showing a failure to ensure that the remote operators were properly selected, trained and entrusted with the operation of MASS. However, in practice it might well be hard to establish such claim.

Given this lacuna in the existing liability regime, which would have the potential to leave an individual uncompensated in the case of a remote operator who was insolvent, hard to locate or uninsured, there may be good arguments for an international regime establishing strict liability in the owner in those circumstances.[54] Undoubtedly, such a regime would promote certainty in the law, as well as protect the rights of claimants. Furthermore, it could serve a deterrent purpose, by encouraging better practices on the part of shipowners.[55] However, it should not be overlooked that such a proposal would be contrary to the provisions of the Convention for the Unification of Certain Rules of Law with Respect of Collisions between Vessels 1910 (the Brussels Collision Convention), which specifically requires liability to be dependent on fault. Understandably, such break from tradition would be resisted by the industry and perhaps by states serving the interests of shipowners, and this could have the potential to impede the universality of the regime governing collision liabilities.

A way around this, it is argued, could be to adopt a provision that would establish a form of deemed vicarious liability for the shipowner in cases where MASS remote operators would be independent contractors. A similar provision has been adopted by the Brussels Collision Convention in the context of pilotage. In particular, Article 5 of the Convention stipulates that "the liability imposed by the preceding Articles attaches in cases where the collision is caused by the fault of a pilot, even when the pilot is carried by compulsion of

53 As seen earlier, MASS remote operators would be under a duty to exercise reasonable skill and care for the safe navigation of those vessels.

54 It may be worth noting here that, in October 2020, the European Parliament proposed to the Commission a regime establishing strict liability for operators of high-risk artificial intelligence (AI) systems in cases "where a physical or virtual activity, device or process driven by an AI-system has caused harm or damage to the life, health, physical integrity of a natural person, to the property of a natural or legal person or has caused significant immaterial harm resulting in a verifiable economic loss". For more details on this proposal, see S. Baughen, "EU Parliament Suggests Civil Liability Regulation for Artificial Intelligence. Possible Collisions Ahead with the IMO Civil Liability Collisions?" (IISTL blog, 20 November 2020) at https://iistl.blog/2020/11/20/eu-parliament-suggests-civil-liability-regulation-for-artificial-intelligence-possible-collisions-ahead-with-imo-civil-liability-collisions/ (accessed 20 January 2021).

55 For more details on the advantages of a strict liability regime, see B Soyer, "Autonomous Vessels and Third-Party Liabilities: The Elephant in the Room" in B Soyer and A Tettenborn (eds), *New Technologies, Artificial Intelligence and Shipping Law in the 21st Century* (Informa Law from Routledge, 2020) 107 to 110. It may be worth noting here, however, that Professor Soyer argues in favour of an international regime that establishes a strict liability regime for the responsible party in cases where the vessel is autonomously operated. However, some of his justification can still be relevant in the present context.

law". In this way, the regime governing civil liabilities in respect of collisions involving remotely controlled ship operations would remain 'fault based'. Meanwhile the shipowner would not be able to escape their part of the liability by arguing that remote operators were acting as independent contractors.

9.6 Conclusion

This chapter has explored a line of legal questions posed by the introduction of remotely controlled ship operations. First, it has been argued that MASS remote operators cannot be considered seafarers. This is not only because remote control centres will be based ashore, but also because the very nature of the work carried out by those centres will not justify the provision of special protection to those individuals involved in the navigation of MASS.

Instead, it has been submitted, MASS remote operators should be considered as a new type of maritime professionals, the training and certification standards of which require further regulation. How this will be made is yet to be seen. The prospect of a new international legal instrument which would supplement the STCW Convention in the context of remotely controlled ship operations has been discussed. The proposed convention on the standards of training and certification for MASS remote operators can follow the structure of other conventions adopted by the IMO, as for the substantive requirements for qualification and training, lessons can be learned from other areas, in which individuals have to be qualified in navigation as well as being properly trained to work with IT and advanced technological systems. The IMO guidelines for VTS services may be instructive in this regard.

Furthermore, it has been argued that remotely controlled ship operations have the potential to challenge traditional approaches to maritime liability, both criminal and civil, in respect of collisions, taking into account the new role that MASS remote operators are expected to play in the operation of those vessels. Where a hybrid method of on board and remote operational control is used, the blurred lines of the sphere of responsibility between the master and remote operators can raise questions as to who should be criminally liable when a collision occurs or who should compensate a third party for physical damage to property and/or person caused by such accident. One option, it has been submitted, might be to retain liability on the master. However, this would have the potential to stress the already pressing issue of criminalisation of seafarers. Other options might include apportioning liability based on specific tasks and functions performed by the master and remote operators or recognising concurrent liability on the master and remote operators. Where, on the other hand, there are no seafarers available on board, it is only logical that any responsibility for the navigation of those ships would be transferred to MASS remote operators, although coastal states might raise concerns against such possibility given the relevant enforcement difficulties.

Finally, it has been argued that the possibility of MASS remote operators being engaged as independent contractors, in particular where remote control centres would be expected to handle multiple ships, might render the current fault-based liability regime unsuitable, as it would allow shipowners to dispute liability on their part, either because the doctrine of vicarious liability would not be applicable or because a claim for systemic negligence would not be proven (take, for example, a situation where the shipowner has exercised reasonable skill and care to adopt a safe system of work or a situation where the shipowner has exercised due care in selecting a properly trained and skilled remote operator). A way around this would be to establish an international regime that provided for strict liability on the part of the shipowner in cases where MASS was controlled by remote operators. However, this might not be welcome by the industry and shipowning states. Alternatively, a provision establishing a form of deemed vicarious liability for the shipowner in cases where MASS remote operators would be independent contractors could be adopted. It has been argued that the latter approach is preferable, as it would provide a more balanced solution.

CHAPTER 10

Shipping and climate change

*Simon Baughen**

10.1 Setting the scene

The world faces an "existential threat" in climate change – as acknowledged by judges in both the UK and the US.[1] The Intergovernmental Panel on Climate Change (IPCC) reports show that: to stabilise at 2°C over pre-industrial levels (taking 1850 as the baseline) by 2100 there must be no more than a concentration of 450 parts per million (ppm) of CO_2; for 1.5°C, the new target set by IPCC in 2018, the concentration must be no more than 430 ppm. The world is currently at about 415 ppm mean concentration, rising about 2.9 ppm annually, with a slight reduction in increase in 2020 due to the effect of coronavirus. The maths shows that we shall hit 430 ppm in five to six years' time and 450 ppm in about 13 years' time. This is confirmed by the figures in the 2018 IPCC SR-1.5 report quoted by Greta Thunberg in her Montreal speech of 27 September 2019:

> In the IPCC [Intergovernmental Panel on Climate Change] SR-1.5 report that came out last year . . . it says on page 108 in chapter 2 [table 2.2] that to have a 67 per cent chance of staying below 1.5 degrees of global temperature rise (the best odds given by the IPCC) the world had 420 gigatonnes of CO_2 left to emit back on Jan. 1, 2018. And today, that figure is already down to less than 350 gigatonnes.

Once in the atmosphere CO_2 remains there for a very long time, so achievement of either of the IPCC targets will require extracting CO_2 from the atmosphere through significant reforestation and the use of as yet untested technologies for direct air extraction (DAC). It is worth remembering that with all the talk of nations becoming "carbon neutral" by 2050, off-setting remaining national emissions to achieve this, countries will also have to participate in a global effort to remove CO_2 from the atmosphere to enable the 1.5 degrees target to be met.

The 2015 Paris Agreement by Art. 2 commits the parties to limit the global temperature rise above pre-industrial levels "to well below 2°C above pre-industrial levels and to pursue efforts to limit the temperature increase to 1.5°C above pre-industrial levels". Article 4.1 then goes on to provide that the long-term emissions reduction goal the agreement expresses can be summarised as aiming for "net zero" in the second half of this century as a way of keeping maximum global temperature rise well below 2°C/1.5°C. At the time of the agreement,

* **Professor of Shipping Law, Institute of International Shipping and Trade Law, Swansea University**

1 *Plan B Earth v Secretary of State for Business, Energy and International Strategy, and the Committee on Climate Change* [2018] EWHC 1892 (Admin) at [48]; *Juliana v US 9th Circuit* 17.1.2020, in which Circuit Court Judge Andrew Hurwitz also referenced "The Eve of Destruction" a popular song from the mid-1960s. While acknowledging that "The plaintiffs in this case have presented compelling evidence that climate change has brought that eve nearer" he nonetheless found against the plaintiffs, as did Supperstone J in the *Plan B* case in the UK.

DOI: 10.4324/9781003155195-10

it was estimated that carrying on with business as usual would lead to an increase in temperature of 4.5–5°C over pre-industrial levels by 2100. To put that in context, it has been stated that the difference between 2 degrees of warming and 4°C of warming can be expressed in a single word: civilisation.

Maritime transport is the most efficient way of shipping goods – better than by truck, rail or air. It is responsible for moving more than 80% of traded goods (by weight). However, international shipping in 2018 accounted for around 2.8% of global CO_2 emissions – about the annual contribution made by Germany or Brazil – and is projected to grow by between 50% and 250% by 2050 if no action is taken. By 2050 it could be producing as much as 17% of global emissions, if not mitigated.

A report by the International Council on Clean Transport in 2017 summarises the substantial, and increasing, contribution of international shipping to the emission of greenhouse gases.

> Shipping GHG emissions are increasing despite improvements in operational efficiency for many ship classes. Increasing emissions are being driven by rising demand for shipping and the associated consumption of fossil fuels.
>
> - Emissions are concentrated in a handful of ship classes and flag states. Just three ship classes (container ships, bulk carriers, and oil tankers) account for 55% of CO_2 emissions. Similarly, six flag states (Panama, China, Liberia, Marshall Islands, Singapore, and Malta) account for 52% of CO_2 emissions.
> - Black carbon is a major contributor to shipping's climate impacts. On a 20-year timescale, BC accounts for 21% of CO_2-eq emissions from ships.
> - The biggest ships are speeding up and emitting more GHGs. Unlike most ships, the largest container and oil tankers sped up between 2013 and 2015 and became less efficient, emitting more CO_2/dwt-nm in 2015 than they did in 2013. As more ships follow their lead, shipping efficiency will drop and ship emissions will continue to rise.
> - Absolute reductions in ship emissions will require concerted action to improve the energy efficiency of shipping and to develop and deploy alternative fuel and propulsion concepts. The only way to reduce emissions from ships without constraining demand is to substantially reduce the amount of CO_2 and CO_2-eq emitted per unit of transport supply.[2]

The reference to black carbon is a timely reminder that CO_2 is not the only shipping emission that is contributing to global warming. Ironically, the International Maritime Organization's (IMO) recent Sulphur Cap and the use of low sulphur heavy fuel oil will lead to an increase in black carbon. There is also nitrogen dioxide and methane, the latter a far more potent warming gas than CO_2 although one that does not linger in the global atmosphere for as long as CO_2. Awareness of the global warming impact of methane is critical in assessing the viability of alternative fuels, particularly as regards liquefied natural gas (LNG) which is currently being advocated. This will be discussed further later in the chapter.

However, along with international aviation, international shipping is not covered by either the 1997 Kyoto Protocol or the 2015 Paris Agreement. It is not covered by the UK's Climate Change Act 2008 either, and although the Committee on Climate Change recommended in autumn 2019 that it be included, the government rejected this.

2 N. Olmer, B. Comer, B. Roy, X. Mao and D. Rutherford, "Greenhouse Gas Emissions from Global Shipping, 2013–2015", International Council on Clean Transportation report published in October 2017 at 25.

A major problem with international shipping is that it is international and multi-faceted in its diversity of actors. This makes it almost impossible to attribute its emission to any one state. Take, as an example, a vessel owned by a Danish company, flying a Marshall Islands flag, chartered by a French shipping company, crewed mainly by Russians and Malaysians, that loads containers in Shanghai, offloading and then loading containers in Singapore, as well as in ports in many countries in Northern Europe. The emissions on that voyage involve up to ten different states.

Under the Kyoto Protocol 1997 *domestic* aviation and shipping emissions are included in national targets for developed countries (art 1), For *international* aviation and shipping emissions the Kyoto Protocol (art 2.2) mandated its parties to work through the IMO to reduce emissions from international shipping, and through the International Civil Aviation Organization (ICAO) to reduce emissions from international aviation. Prior to the Kyoto Protocol's adoption in 1997, the Convention's Subsidiary Body on Technical Advice (SBSTA) was requested to examine the allocation and control of emissions from international bunker fuels and considered five options which were left open as no agreement could be reached on their importance.

- No allocation;
- Allocation to the country where the bunker fuel is sold;
- Allocation to the country of the transporting company; the country of registration of the aircraft/vessel, or the country of the operator;
- Allocation to the country of departure or destination of the aircraft/vessel (including sharing emission between them);
- Allocation to the country of departure or destination of the passenger/cargo (including sharing emissions).

Therefore, it is all down to the IMO. So, what has the IMO done? One of the strands of the IMO's GHG strategy is improving energy efficiency via technological measures. Under Annex VI of MARPOL[3] ships built after 1 January 2013 must comply with a minimum energy efficiency level: the Energy Efficiency Design Index (EEDI). The EEDI measures the CO2 emitted (g/tonne-mile) based on ship design and engine performance data. The EEDI level is tightened incrementally every five years with an initial CO2 reduction level of 10% for the first phase (2015–2020), 20% for the second phase (2020–2025) and a 30% reduction mandated from 2025 to 2030.

The IMO has also introduced reporting requirements on CO2 emissions, made amendments to the Ship Energy Efficiency Management Plan (SEEMP) and provided for guidelines for fuel-consumption data collection. These were adopted by the IMO's Marine Environment Protection Committee (MEPC70) on 28 October 2016 as amendments to chapter 4 of Annex VI of MARPOL, adding a new Regulation 22A on Collection and reporting of ship fuel oil consumption data and new appendices covering information to be submitted to the IMO Ship Fuel Oil Consumption Database.

But there is another player in the game – The European Union. Its Monitoring, Reporting and Verification Regulation 2015/757 requires reporting of three items in respect of all merchant vessels above 5,000 grt, calling at ports in the EU and the EEA: actual cargo

3 Regulations 19, 20 and 21 of chapter 4 of MARPOL Annex VI.

carried onboard; fuel consumed; and CO_2 emitted. It differs from the IMO regime in three aspects. First, the IMO requires reporting only of fuel consumed, not of CO_2 emitted.[4] Second, the Regulation requires mrv calculations to be verified by an accredited verifier and sent to a central European Commission (EC) database, likely to be managed by the European Maritime Safety Agency (EMSA), whereas the IMO requires verification by the flagship administration, under national procedures. Third, the Regulation provides for this information to be publicly available, as opposed to the IMO where raw data is restricted to the IMO and flag states and aggregated anonymised data is then made public. Fourth, the IMO does not require reporting of cargo carried.

The EU monitoring scheme started on 1 January 2018, with the first reporting deadline arriving on 30 April 2019. The IMO scheme started on 1 January 2019. A similar parallel regulatory dance can be seen with regulation of the sulphur content of fuel oil. The IMO sulphur limits were initially 3.5% and 0.10% for ships operating within emission control areas (ECAs). A global cap of 0.5% was set to be introduced between 2020 and 2025. The EU's Sulphur Directive has the same limit for vessels within ECAs but also applies it for ships at berth in any EU port. From 2020 a cap of 0.50% would apply within EU waters, defined as the Economic Exclusive Zone (EEZ). In October 2016 at the 70th meeting of the Marine Environment Protection Committee (MEPC70) the IMO decided to introduce a global cap of 0.50% on 1 January 2020 – and much of 2019 was spent discussing its implications for shipowners and charterers.

And what about reducing GHG emissions from international shipping? The impetus for IMO action came, yet again, from the EU. The EU Parliament proposed on 10 November 2017 to add shipping to the revised EU Emissions Trading System (ETS) from 2023 if IMO progress in a CO_2 strategy was considered insufficient. That was followed by the IMO's Marine Environment Protection Committee (MEPC) 72 in April 2018 in which the IMO stated its ambition to reduce greenhouse gas emissions from shipping by 50% over 2008 levels by 2050, while, at the same time, pursuing efforts towards phasing them out entirely. The strategy sets a carbon intensity reduction target – the amount of emissions relative to each tonne of shipping cargo – of at least 40% by 2030, rising to 70% by 2050. The strategy includes a specific reference to "a pathway of CO_2 emissions reduction consistent with the Paris Agreement temperature goals".

In May 2019 MEPC 74 approved, for adoption at the next session in April 2020, amendments to MARPOL Annex VI to significantly strengthen the EEDI "phase 3" requirements. The draft amendments bring forward the entry into effect date of phase 3 to 2022, from 2025, for several ship types, including gas carriers, general cargo ships and LNG carriers. This means that new ships built from that date must be significantly more energy efficient

4 CO_2 emissions are calculated by applying the following formula: Fuel consumption × emission factor. "Fuel consumption includes fuel consumed by main engines, auxiliary engines, gas turbines, boilers and inert gas generators, and fuel consumption within ports at berth is to be calculated separately. In principle, default values for emission factors of fuels shall be used unless the company decides to use data on fuel quality set out in the Bunker Fuel Delivery Notes (BDN) and used for demonstrating compliance with applicable regulations of sulphur emissions. Those default values for emission factors shall be based on the latest available values of the Intergovernmental Panel for Climate Change (IPCC). Those values can be derived from Annex VI to Commission Regulation (EU) No. 601/2012 (1). Appropriate emission factors shall be applied in respect of biofuels and alternative non-fossil fuels" at www.verifavia-shipping.com/shipping-carbon-emissions-verification/faq-which-methods-can-be-used-to-calculate-fuel-consumption-26.php (accessed March 2021).

than the baseline. For container ships, the EEDI reduction rate is enhanced, significantly for larger ship sizes.

The terms of reference for the Fourth IMO GHG Study were set out as follows. There should be an inventory of current global emissions of GHGs and relevant substances emitted from ships of 100 GT and above engaged in international voyages. The inventory should include: total annual GHG emission series from 2012 to 2018, or as far as statistical data are available; estimates of carbon intensity (estimates of world fleet's CO_2 emissions per transport work, from 2012 to 2018, or as far as statistical data are available); possible estimates of carbon intensity of international shipping for the year 2008 (the baseline year for the levels of ambition identified in the Initial Strategy); scenarios for future international shipping emissions 2018–2050.

The MEPC adopted resolution MEPC.323(74) on Invitation to Member States to encourage voluntary cooperation between the port and shipping sectors to contribute to reducing GHG emissions from ships. This could include regulatory, technical, operational and economic actions, such as the provision of: Onshore Power Supply (preferably from renewable sources); safe and efficient bunkering of alternative low-carbon and zero-carbon fuels; incentives promoting sustainable low-carbon and zero-carbon shipping; and support for the optimisation of port calls including facilitation of just-in-time arrival of ships. The procedure identifies four steps:

Step 1: initial impact assessment, to be submitted as part of the initial proposal to the Committee for candidate measures;
Step 2: submission of commenting document(s), if any;
Step 3: comprehensive response, if requested by commenting document(s); and
Step 4: comprehensive impact assessment, if required by the MEPC. Shipping and Climate Change.

The MEPC agreed to establish a voluntary multi-donor trust fund ("GHG TC-Trust Fund") to provide a dedicated source of financial support for technical cooperation and capacity-building activities to support the implementation of the Initial IMO Strategy on reduction of GHG emissions from ships. The MEPC discussed various candidate short-term measures, including strengthening the energy efficiency requirements for existing ships, speed and other technical and operational measures. In view of the vast number of proposals, the working group focused on how to consider, organise and streamline proposals on candidate short-term measures. The intersessional working group session will further consider candidate short-term measures, including concrete proposals to improve the operational energy efficiency of existing ships.

One of the candidate short-term measures is speed regulation. In April 2019 the French delegation had proposed mandatory slow steaming, for international shipping with the exception of container ships. But, the response from the ship-owning community was a clear "Non".[5] Slow steaming would lead to longer voyage times, leading to higher operating costs, insurance and employment expenses that come with operating a greater number of ships at any given time. Shipowners would be encouraged to buy in extra tonnage which would delay the introduction of new energy efficient vessels. As an example of the obstacles to any short-term agreement on how to peak greenhouse gas emissions in the near term, no agreement could

5 A tweet from The Wall Street Journal said, "At this point, it's basically dead", in reference to speed limits. Others were less certain – according to one close observer of the IMO, "I think it is too soon to definitively rule anything in or out. The discussion will continue."

even be reached on which of several possible short-term measures to discuss first, with Brazil, Saudi Arabia and the US objecting even to the word "prioritisation".

The MEPC also considered concrete proposals on candidate mid-/long-term measures, in particular measures aimed at encouraging the uptake of alternative low-carbon and zero-carbon fuels. One such measure would be mandatory slow steaming, as the most effective way for an immediate reduction in GHG emissions.

Currently on the table before the IMO are two proposals for a bunker levy. The first, at the start of 2020, by shipowners was for a small levy of $2 per tonne to establish a fund to research into alternative fuels. The second was a on 25 September, when the major charterer, Trafigura, submitted a proposal to the IMO for a partial "feebate" system to decarbonise global shipping. Trafigura's press release states:

> We propose a self-financing system where a levy is charged on the use of fuels with a CO_2-equivalent intensity above an agreed benchmark level, and a subsidy is provided for fuels with a CO_2-equivalent profile below that level. It is now time to put a price on carbon emissions in the shipping industry. Our own in-depth analysis and commissioned independent research indicates that the levy should be between $250–$300 per tonne of CO_2-equivalent. While primarily bridging the cost gap between carbon intensive and low or zero carbon fuels, this partial "feebate" would also raise billions of dollars for research into alternative fuels and could help assist small island developing states and other developing countries mitigate the impact of climate change.

Further measures to reduce GHG emissions from ships were discussed when the IMO's Marine Environment Protection Committee (MEPC) met between 16 and 20 November 2020 to discuss measures to reduce further GHG from shipping. The MEPC was expected to adopt amendments to the International Convention for the Prevention of Pollution from Ships (MARPOL) to significantly strengthen the "phase 3" requirements of the EEDI – meaning that new ships built from 2022 will have to be significantly more energy efficient. Those amendments were approved at the previous session of the Committee (MEPC 74) in May 2019. The MEPC also discussed two further energy efficiency requirements comprising draft amendments which were agreed by IMO's Intersessional Working Group on Reduction of GHG Emissions from Ships (ISWG-GHG 7) in October, and would also apply to existing ships:

a new Energy Efficiency Existing Ship Index (EEXI) for all ships;
an annual operational carbon intensity indicator (CII) and its rating, which would apply to ships of 5,000 gross tonnage and above.

If approved at this session of the Committee, they could then be put forward for adoption at the subsequent MEPC 76 session, to be held in June 2021. Under MARPOL, amendments can enter into force after a minimum 16 months following adoption.

Finally, there is the final report of the Fourth IMO Greenhouse Gas Study, released on 4 August 2020.[6] This highlights the urgency of the need to decarbonise shipping and found that total GHG emissions from maritime shipping rose about 10% from 2012 to 2018. In particular:

> GHG emissions – including CO_2, CH_4 and N_2O, expressed in CO_{2e} – of total shipping (international, domestic and fishing) have increased from 977 million tonnes in 2012 to 1,076 million tonnes in 2018 (9.6% increase). In 2012, 962 million tonnes were

6 www.imo.org/en/OurWork/Environment/Pages/Fourth-IMO-Greenhouse-Gas-Study-2020.aspx (accessed 5 April 2021).

CO_2 emissions, while in 2018 this amount grew 9.3% to 1,056 million tonnes of CO_2 emissions.

The share of shipping emissions in global anthropogenic GHG emissions has increased from 2.76% in 2012 to 2.89% in 2018.

Carbon intensity (fleet's CO_2 emissions per transport work) has improved for international shipping as a whole, as well as for most ship types. For instance, in 2018, the overall average carbon intensity across international shipping was about 30% lower than in 2008, as measured by the IMO's Energy Efficiency Operational Indicator (EEOI). However, more than half of the fuel efficiency improvements have been achieved before 2012. The pace of carbon intensity reduction has slowed since 2015, with average annual percentage changes of only 1–2%.

GHG emissions are projected to increase from about 90% of 2008 emissions in 2018 to 90–130% of 2008 emissions by 2050 for a range of six plausible long-term economic and energy scenarios.

150% increase in methane emissions from 2012 to 2018. The increase in CH_4 emissions was largely due to a surge in the number of ships fueled by liquefied natural gas (LNG), many of which produce high Methane is a powerful warming agent, with the global warming potential over a 20-year period (GWP_{20}) of 86.

Black Carbon emissions increased approximately 12% from 2012 to 2018. The IMO is scheduled to agree on black carbon regulations in 2021.

10.2 Possible short-term measures

A report prepared for the European Commission in April 2019 by CE Delft[7] analysed the effect of various possible short-term measures before the IMO to be agreed by 2023, with the purpose of meeting its 2030 target of reducing CO_2 emissions per transport work, as an average across international shipping, by at least 40% by 2030 compared to 2008. In order to achieve a 40% carbon intensity reduction on 2008 carbon intensity, the report estimated that a CO_2 emissions reduction of approximately 21% in 2030 relative to the business as usual (BAU) would be required. The measures could be grouped into three categories.

(1) Measures that can help remove barriers to the implementation of cost-effective technologies or operational practices: strengthening the Ship Energy Efficiency Management Plan (SEEMP); mandatory goal-setting with a self-chosen goal; strengthening the SEEMP; mandatory periodic efficiency assessment without a goal; developing a standard for ship-shore communication for voluntary use; developing a standard for port incentive schemes for voluntary use; creating a framework for incentivising the uptake of renewable fuels.

(2) Measures that mandate ships to improve their technical or design efficiency: strengthening the EEDI for new ships; applying the EEDI to existing ships; strengthening the SEEMP – mandatory retrofits of cost-effective technologies; existing fleet improvement programme.

7 "Study on Methods and Considerations for the Determination of Greenhouse Gas Emission Reduction Targets for International shipping. Final Report: Short-term Measures." Available at www.ucl.ac.uk/bartlett/energy/news/2019/may/shipping-team-publishes-paper-how-greenhouse-gas-reduction-targets-can-realistically (accessed 22 March 2021).

(3) Measures that mandate operational carbon intensity improvements:[8] setting mandatory operational efficiency standards; speed regulation.

The impact of the first category of measures had the least effect of reduction as against BAU. The second category had more impact but neither measure would enable the 2030 target to be achieved. However, that would be possible with the third category which would achieve and surpass the 21% reduction required. Speed reduction capping average speed at 20% below the 2012 level would give a reduction of between 24–34% while operational efficiency standards with annual efficiency ratio (AER) 40% below the 2008 level would give a 21% reduction which would rise to 43% with operational efficiency standards with AER 60% below 2008.

10.3 Alternative fuels

The IMO's strategy also recognises that in order to meet the 2050 Level of Ambition, the global introduction of alternative fuels and/or energy sources will be required. In the longer term, which, *pace* the IMO and its 2050 ambition, means within the next 15 years in the context of global warming, shipowners will have to adjust to the use of alternative fuels. LNG, hydrogen, electric, nuclear, biofuel,[9] ammonia, and even a partial return to sail,[10] have all been touted.

---

8 The report notes at 2.7.1 that: "Several operational efficiency metrics have been proposed and discussed at the IMO. The most recent proposals are the following:

1 Energy Efficiency per Service Hour (EESH) (MEPC 65/4/19).
2 Annual Efficiency Ratio (AER) (MEPC 66/4/6, MEPC 67/5/4).
3 Individual Ship Performance Indicator (ISPI) (MEPC 66/4/6).
4 Fuel Oil Reduction Strategy (FORS) (MEPC 66/4/6)".

All indicators also incentivise speed reduction, although not all to the same extent.

9 BBC News, "Biofuel Danish firm claims first biofuel commercial sea voyage". Available at www.bbc.co.uk/news/business-46394339?utm_source=Greenhouse+Morning+News&utm_campaign=acf3231b52-Greenhouse_Morning_News_November30th_2018&utm_medium=email&utm_term=0_e40c447c1a-acf3231b52-123999121 (accessed March 2021).

> "When burnt, the fuel produces carbon dioxide. But the firm says this is offset by the fact that the fuel is made from organic material, which has absorbed a similar amount of the gas at a previous stage." However, this view of carbon offsetting from biofuels is rebutted by a blog in The Conversation of 5.10.16 by John DeCicco who states: "It turns out that once all the emissions associated with growing feedstock crops and manufacturing biofuel are factored in, biofuels actually increase CO2 emissions rather than reducing them . . . There are two broad strategies for mitigating CO2 emissions from transportation fuels. First, we can reduce emissions by improving vehicle efficiency, limiting miles traveled or substituting truly carbon-free fuels such as electricity or hydrogen. Second, we can remove CO2 from the atmosphere more rapidly than ecosystems are absorbing it now. Strategies for 'recarbonizing the biosphere' include reforestation and afforestation, rebuilding soil carbon and restoring other carbon-rich ecosystems such as wetlands and grasslands." Available at https://theconversation.com/biofuels-turn-out-to-be-a-climate-mistake-heres-why-64463 (accessed March 2021).

10 At the end of 2018 Renault announced that it was partnering with Neoline to build two experimental roll-on/roll-off car carriers powered by sails. Each will be 446 feet long and carry more than 45,000 square feet of sails. Speed would be reduced to 11 knots. See https://cleantechnica.com/2018/12/24/renault-will-use-sails-to-cut-emissions-on-trans-atlantic-routes/ (accessed March 2021).

In November 2020 it was announced that Wallenius Marine was working with the Swedish government on a proposed sail-powered car carrier that could cut CO_2 emissions by 90%, though at the cost of a reduction in speed from 17 knots to 10 knots, adding an extra five days to an Atlantic crossing. Five telescopic wing-shaped elements stand 260 ft above the deck. They could rotate 360 degrees without touching each other, or retract to 195 ft in rough weather, and where it was necessary to allow the ship to pass under bridges.

www.greencarreports.com/news/1130302_wind-powered-transatlantic-car-carrier-could-cut-carbon-emissions-by-90 (accessed March 2021).

Other possibilities for reducing shipping's carbon footprint are marine geo-engineering,[11] ship redesign, "cold ironing" at ports, more effective port scheduling. There is also electric propulsion by batteries of the sort scheduled for the Norwegian prototype unmanned coastal container vessel, the "Yara Birkeland", although battery power is only feasible for short journeys, such as the coastal transits in Norwegian waters proposed by that vessel.

In considering these options, account has to be taken of the full life-cycle of the replacement fuel. If it is from hydrogen that has to be from "green hydrogen" which is produced by renewable electricity, similarly for "cold ironing" and the use of batteries. Furthermore, a report by Faig Abbasov, and others, of NGO Transport and Environment "Roadmap to Decarbonising European Shipping. Nov 2018"[12] points out that the complete decarbonisation of EU-related shipping in 2050 would require 11–53% additional renewable electricity generation across the EU28 (now 27) over the 2015 levels. This range is estimated on the assumption that EU maritime emissions will grow by around 50% between 2010 and 2050, taking into account the deployment of a range of short and mid-term measures, EEDI & SEEMP, speed reduction and wind propulsion.

The recent ABS Future Fuels LinkedIn Survey of shipowners showed that LNG, which already fuels some container vessels, and hydrogen are viewed as frontrunners to be the fuels that will help shipping reach the 2050 IMO decarbonisation ambitions.[13] Out front was LNG with 47% of the vote, followed by hydrogen at 40%, ammonia at 8% and methanol at 5%. With regard to the 2030 targets LNG polled 64%, hydrogen 22%, and 7% each for ammonia and methanol. However, there are severe drawbacks with LNG as a replacement fuel as noted in the International Transport Forum's 2018 report, as follows:

> Its advantage is the ability to reduce SO_X emissions and particulate matter by almost 100% compared to conventional fuel oil (IMO 2016), but the major disadvantage is "methane slip". Its handling and combustion involves the release of methane, a very potent GHG with global warming potential 28 times higher than CO_2 over a period of 100 years and 84 times higher over a 20 years period (Anderson et al., 2015). Methane slip can also occur during bunkering phase as well as upstream in the fuel production, processing and transmission. The reduction in CO_2 emissions is estimated to be about 25%.[14]

This analysis was reiterated in a paper published in January 2020 which analysed upstream emissions, combustion emissions, and unburned methane (methane slip), using 100-year and 20-year global warming potentials (GWPs), comparing the life-cycle GHG emissions of LNG, marine gas oil (MGO), very low sulphur fuel oil, and heavy fuel oil when used in engines suitable for international shipping, including cruise ships.[15] Over a 100-year timeframe, the maximum life-cycle GHG benefit of LNG was a 15% reduction compared with MGO, and this was only if ships used a high-pressure injection dual fuel (HPDF) engine and upstream methane emissions were well-controlled. Using a 20-year GWP to reflect the

---

11 See www.imo.org/en/MediaCentre/PressBriefings/Pages/04-marinegeoengineeringGESAMP.aspx (accessed March 2021).

12 www.transportenvironment.org/sites/te/files/publications/2018_11_Roadmap_decarbonising_European_shipping.pdf (accessed March 2021).

13 https://news.cision.com/american-bureau-of-shipping/r/lng-and-hydrogen-emerge-as-future-frontrunners-according-to-stakeholder-survey-abs-simplifies-opti,c3217061 (accessed March 2021).

14 www.itf-oecd.org/decarbonising-maritime-transport-2035 (accessed March 2021).

15 N. Pavlenko, B. Comer, Y. Zhou, N. Clark and D. Rutherford D, *The Climate Implications of Using LNG as a Marine Fuel.* https://theicct.org/publications/climate-impacts-LNG-marine-fuel-2020 (accessed March 2021).

urgency of reducing GHGs to meet the climate goals of the IMO, and factoring in higher upstream emissions for all systems and crankcase emissions for low-pressure systems, it concluded that there was no climate benefit from using LNG, regardless of the engine technology.

> HPDF engines using LNG emitted 4% more life-cycle GHG emissions than if they used MGO. The most popular LNG engine technology is low-pressure dual fuel, four-stroke, medium-speed, which is used on at least 300 ships; it is especially popular with LNG fueled cruise ships. Results show this technology emitted 70% more life-cycle GHGs when it used LNG instead of MGO and 82% more than using MGO in a comparable medium-speed diesel (MSD) engine. Given this, we conclude that using LNG does not deliver the emissions reductions required by the IMO's initial GHG strategy, and that using it could actually worsen shipping's climate impacts. Further, continuing to invest in LNG infrastructure on ships and on shore might make it harder to transition to low-carbon and zero-carbon fuels in the future. Investing instead in energy-saving technologies, wind-assisted propulsion, zero-emission fuels, batteries, and fuel cells would deliver both air quality and climate benefits.

The second-placed alternative fuel, hydrogen, is more promising. Hydrogen can be used as fuel in several different ways, that is, in fuel cells; in a dual fuel mixture with conventional diesel fuels (HFO); and lastly as a replacement for HFO for use in combustion machinery. It emits zero CO_2, zero sulphur oxides (SOx) and only negligible amounts of nitrogen oxides (NO_X). Grey, or brown, hydrogen is made using fossil fuels like oil and coal, which emit CO_2 into the air as they combust. The blue variety is made in the same way, but carbon capture technologies prevent CO_2 being released, enabling the captured carbon to be safely stored deep underground or utilised in industrial processes. Green hydrogen is the cleanest variety, producing zero-carbon emissions. It is produced using electrolysis powered by renewable energy, like offshore wind, to produce a clean and sustainable fuel.

A 2017 report by SANDIA examined the viability of scaled-up zero-emission powertrains to meet one voyage or "trip" using energy storage characteristics of the battery system.[16] Three systems were examined: (i) a proton exchange membrane (PEMFC) fuel cell using liquid hydrogen as a cryogenic liquid at about 20°K or -252°C; (ii) a PEMFC fuel cell, as a high-pressure gas at 5,000 psi (350 bar); (iii) a battery. Vessels' physical characteristics were used to estimate a maximum mass and volume limit for the powerplant and fuel. No vessel redesign was considered; all case studies used existing layouts.

The study then sized up each of these three zero-emission powertrains to see how they could meet one voyage or "trip" for various types of vessel, using energy storage characteristics of the battery system. Three systems were examined: (i) a PEMFC fuel cell using liquid hydrogen as a cryogenic liquid at about 20°K or -252°C; (ii) a PEMFC fuel cell, as a high-pressure gas at 5,000 psi (350 bar); (iii) a battery.

The largest vessel considered was the *Emma Maersk* with main engine power of 80.08 MW for a 5005 nm voyage from Tanjung Pelepas to Suez Canal/Port Said, with an average speed of 19.6 knots, and total voyage time of 256 hours. It found that:

> "Despite the large power and energy requirements, the available mass and volume is able to accommodate a hydrogen fuel cell powerplant using liquid hydrogen as the fuel. The fuel cell system with 350 bar hydrogen meets the mass requirements but is more than 3-times the available volume. The battery system is over 15 times too large and massive." In conclusion "both fuel

16 J. J. Minnehan and J. W. Pratt, Practical Application Limits of Fuel Cells and Batteries for Zero Emission Vessels/ At https://energy.sandia.gov/wp-content/uploads/2017/12/SAND2017-12665.pdf (accessed March 2021).

cell systems work in almost every vessel and can be implemented in longer distances than the battery systems. It is only at the longer voyages that the 5000 psi hydrogen powered fuel cell fails to meet the requirements. For the greatest versatility and efficiency, the LH_2 FC system prevails."

Two further points are noted in the conclusion:

> Although both types of powerplants are zero emission on the ship, the energy used to recharge the batteries or make the hydrogen may not be zero emission. Full "well-to-waves" emissions analysis with comparison to conventional diesel technology is needed to understand the true global impact of converting vessels to zero emission.

And,

> However, both fuel cell and batteries produce Direct Current (DC) and require a converter. A converter could occupy a significant amount of space on a marine vessel. This becomes important when designing a battery or fuel cell power plant and its location on a ship. What is the availability of electric motors in the 10s of MW and how do they change the assessment? How does the selection of AC or DC motor change the other equipment required and which solution is better?

A further paper published on 3 March 2020[17] modelled the fuel demand and space requirements of container ships when powered by hydrogen, stored as a cryogenic liquid and used in a fuel cell, instead of heavy fuel oil (HFO) in an internal combustion engine. It then calculated what share of legs and voyages could be completed with hydrogen, given the ships' available space for fuel and propulsion equipment, and evaluated two options; one to sacrifice some cargo space to put more hydrogen fuel on board; the other to allow more frequent refuelling en route. It concluded:

> Our main finding is that 99% of the voyages made along the corridor between Pearl River Delta (PRD) region in China and the San Pedro Bay (SPB) area in the United States can be powered by hydrogen instead of fossil fuels, with only minor changes to ships' fuel capacity or operations. Specifically, this could be achieved by replacing 5% of certain ships' cargo space with more hydrogen fuel, or by adding one additional port call to refuel hydrogen en route. Importantly, we also find that 43% of all voyages could be completed without adding any fuel capacity or extra port calls.

The paper concluded with an important caveat.

> To boost early investment in zero-emission technology and associated bunkering infrastructure, we also need economic assessments of such corridors. This involves not only the cost of technology implementation for the applicable fleet and the profitability of certain types of commodity trading utilizing a given corridor, but also the cost of developing port infrastructure.
>
> Currently, HFO is produced in fossil fuel abundant regions and transported to major bunkering ports. If more ports need to offer bunkering services to support a zero emission fleet, a distributed hydrogen production and delivery network would be needed. Life-cycle assessments of alternative marine fuels would also be needed to ensure that they are low carbon throughout their production cycle.

However, use of fuel cells on large container vessels will require significant scaling-up from existing systems. "[T] largest fuel cell systems today produce only 1–5 MW", say DFDS who are currently working in a partnership to develop a 100% hydrogen powered ferry for DFDS' Oslo–Frederikshavn–Copenhagen route to be running by 2027. The ferry will be powered by electricity from a hydrogen fuel cell system that emits only water and can

17 https://theicct.org/publications/zero-emission-container-corridor-hydrogen-2020 (accessed March 2021).

produce up to 23 MW to propel the ferry. It is estimated that between 60–100 MW is needed to power container ships. On a larger scale, a blog entry on 14 April 2020 by Loz Blain noted that ABB and Hydrogène de France were teaming up to build enormous hydrogen fuel cell powertrains for large marine vessels.[18] The new agreement would see the development of a "megawatt-scale" powertrain for marine vessels, using HDF's large-scale manufacturing capabilities to get the thing built. The design would be done in conjunction with Ballard Power Systems, specialists in proton exchange membrane (PEM) fuel cells. There is currently no indication of how big ABB plans to go with its first powertrains, but there is an indication that they might be able to build something for use in a hybrid arrangement on larger ships, where hydrogen could be used to "support auxiliary energy requirements of larger vessels".

However, it may soon be possible for fuel cells to be retrofitted to ships and used to provide additional fuel to conventional oil or diesel engines. At Swansea University, Charlie Dunnill at Energy Safety Research Institute (ESRI) is working on the Desire project whereby waste energy and wind power are used to split water into hydrogen and oxygen which are then both added to the air intake of the oil or diesel engine. This should improve the burn, lower the emissions and increase the power per unit of fuel, thus lowering the costs.

The use of sea water to create hydrogen with the help of solar panels is also currently being explored by Toyota in the *Energy Observer*, a former catamaran.[19] When the Energy Observer began its journey in 2017, it could only produce hydrogen while stopped. That changed with the addition of the Oceanwings 12 m sails that improved the efficiency of the *Energy Observer* from 18% to 42%, to the point where it can now produce hydrogen even while sailing. The *Energy Observer* is equipped with a reverse-osmosis desalination system with several levels. When two volumes of water – one salty and the other not – are put together, natural movement is created: the fresh water is attracted to the salty water. On the *Energy Observer*, this process instantly consumes 250 W to produce 90 litres of drinking water, 30 litres of which are then treated again to be used by the electrolyser. One litre of fresh water produces 100 g of hydrogen gas, which becomes water again when the fuel cells make the conversion into electricity.

At joint third place, a small percentage indicated a preference for ammonia and methanol. Ammonia has the advantage that it emits zero CO_2 and SO_2, and close to zero NO_2. When used in a fuel cell it has the advantage of requiring less cargo space that hydrogen. However, it is an extremely toxic substance that militates against its uses as a fuel source. Methanol is now mostly produced from natural gas and has a CO_2 emissions reduction potential of around 25% compared to HFO, with an emission reduction potential of 99% for SO_x, 60% for NO_x and 95% for particle matter (PM). Methanol is available in large quantities and can be made out of a wide number of resources. Methanol can be produced from renewable energy resources, such as CO_2 capture, industrial waste, municipal waste or biomass which significantly reduces its greenhouse impact. A significant advantage is that its use requires only minor modifications to existing combustion engines for ships and bunkering infrastructure. However, the CO_2 emissions reduction potential is low compared to other alternative fuels.

There are also biofuels which are estimated to be able to reduce CO_2 emissions between 25% and 100% for very good biofuels, subject to the important caveat that, excluding biofuels requiring conversion of farm land or forests, the IEA estimate that total biofuels supply can

18 https://newatlas.com/marine/hydrogen-ships-fuel-cell-marine-abb/ (accessed March 2021).

19 https://newsroom.toyota.eu/toyota-develops-specially-designed-fuel-cell-system-for-energy-observers-2020-tour/ (accessed March 2021).

only cover about 15% of total demand.[20] Biofuels were considered in an ICCT report in September 2020[21] which concluded that biodiesels from fats oils and greases (FOGs) would be the most suitable short-term alternative fuel for use in the sector to support the IMO's initial GHG strategy. Use of such fuels would, however, be greatly limited owing to the low supply of waste FOGs in conjunction with high competition from other transport sectors. The report saw Fischer Tropsch (FT) diesel[22] from lignocellulosic biomass as providing greater GHG reductions and as having a much higher feedstock availability, making it a better long-term bet; but its near-term use was limited by its high costs and lower technological readiness.

10.4 Ship financing: the Poseidon Principles

A current "green" financial initiative for shipping is the Poseidon Principles promulgated on 18 June 2019, when 11 banks with collectively over $100 bn in assets and representing nearly 20% of the global ship finance market signed a global framework agreement, directed towards meeting the Paris Agreement's target of an increase below 2°C and the IMO target of cutting greenhouse gas emissions from global shipping by 50% by 2050 compared to 2008 levels. The 11 signatory banks will rely on the global Data Collection System (DCS) for Fuel Oil Consumption by ships (IMODCS) when assessing the carbon intensity of their related ship finance, and will work together in promoting responsible ship finance. To overcome the confidentiality of the IMODCS data, the Principles require signatories to use their best efforts to include a standard covenant in each of their new finance agreements requiring shipowners to provide them with their fuel consumption and other relevant data. The wording of the standard covenant has been provided by the secretariat.

A proforma covenant is provided which requires the owner in each calendar year, at the lender's request, to supply all information necessary in order for the lender to comply with its obligations under the Principles in respect of the preceding year, including all ship fuel oil consumption data required to be collected and reported in accordance with Regulation 22A of Annex VI and any Statement of Compliance, in each case relating to the vessel for the preceding calendar year. This is subject to a confidentiality proviso preventing the lender from publicly disclosing the information indicating the identity of the vessel without the owner's consent. It is not clear what remedies would exist if the owners were to fail to comply with the covenant. The obligation is not worded as a condition; and as an innominate term it is difficult to see how a single breach would go to the root of the contract so as to justify termination. Damages would be likely to be nominal, as the lender would not normally suffer loss as a result of the breach.

Lenders are also committed to publicly acknowledging no later than 30 November in each year their status as a signatory, to reporting the overall climate alignment of their related shipping portfolio to the secretariat with supporting information, and to publish that overall climate alignment in relevant institutional reports on an appropriate timeline. Their climate alignment scores are to be published on the Poseidon Principles website by 31 December each year.

20 International Transport Forum, "Decarbonising Maritime Transport Pathways to Zero-Carbon Shipping by 2035", 2018, 33, at www.itf-oecd.org/decarbonising-maritime-transport-2035 (accessed March 2021).

21 Y. Zhou, N. Pavlenko, D. Rutherford, L. Osipova, and B. Comer, *The potential of liquid biofuels in reducing ship emissions* at https://theicct.org/publications/marine-biofuels-sept2020 (accessed March 2021).

22 FT diesel can be synthesised from fossil fuels such as coal and natural gas, or from lignocellulosic biomass, such as forest residue and willow.

10.5 Future unilateral action by the EU?

In October 2014, the EU set a domestic GHG reduction target of at least 40% below 1990 levels by 2030. Shipping is however currently outside this target, with climate change regulation for international shipping being parked in the slow lane in the IMO. That may, however, be about to change over the next two years.

First, there is the "European Green Deal" set out in the Commission's communication of 11 December 2019.[23] The Commission indicated that it would be looking at measures extending its ETS to shipping, and would also look closely at the current tax exemptions including for aviation and maritime fuels and at how best to close any loopholes. It would in addition take action in relation to maritime transport, including to regulating access of the most polluting ships to EU ports and obliging ships in port to use shore-side electricity. On 4 March 2020 the Commission proposed a Regulation establishing the framework for achieving climate neutrality and amending the European Climate Law.[24] Under this, by September 2020, the Commission would review the Union's 2030 climate target in the light of the climate-neutrality objective set out in Article 2(1) of the Law, and explore options for a new 2030 target of 50% to 55% emission reductions compared to 1990.

Second, on 24 January 2020 Green MEP Jutta Paulus, as Rapporteur for the European Parliament's Committee on the Environment, Public Health and Food Safety produced a draft report[25] on the proposal for a Regulation amending the 2015 CO_2 Emissions Regulation (the MRV Regulation)[26] in order to take appropriate account of the global DCS for ship fuel oil consumption data. The MRV Regulation requires all ships above 5,000 GRT to report their annual fuel consumption and associated CO_2 emitted during voyages between EEA ports, between non-EEA ports EEA ports, and those occurring when the ship is in berth in the EEA. The obligation lies on the shipping operator: that is, the registered owner, the bareboat charterer, or the managing company. The first year of compliance was set for 2018 with the first emissions reports released on 30 June 2019.

On 4 February 2019 the Commission proposed amending the MRV to harmonise it with the IMO's DCS scheme. A major change would be that reporting of cargo carried would become voluntary. The other proposed harmonisations were that "Time at sea" should be replaced by the global IMO DCS definition of "hours underway", and the calculation of "distance travelled" should be based on global IMO DCS25 to reduce the administrative burden.

Jutta Paulus' report proposed radically altering the amendments to the MRV which the Commission had put forward a year earlier. Apart from reinstating the mandatory reporting of cargo carried, this report recommended a number of further significant changes.

First, maritime transport should be included in the ETS, on the same basis on which it was subject to the reporting obligations under the MRV, since its inclusion was a more attractive option than a simple carbon tax on bunkers. Such a tax could only be applied to marine fuels sold in European ports, leaving bunkers sold elsewhere untaxed; furthermore,

23 Communication from the Commission to the European Parliament, the European Council, the Council, the European Economic and Social Committee and the Committee of the Regions: the European Green Deal (Brussels, 11.12.2019: COM(2019) 640 final).

24 Regulation (EU) 2018/1999.

25 COM(2019)0038 – C8–0034/2019–2019/0017(COD).

26 Regulation (EU) 2015/757.

it would require revising the EU Energy Tax Directive[27] to remove its ban[28] on taxation of marine fuel sold to ships on EU territory, something that would require unanimity in the Council, rather than the qualified majority vote which was all that was required for revision of the ETS Directive or the MRV Regulation.

The amended regulation should cover GHG emissions taking place during voyages from a port of call outside the EU to one under the jurisdiction of a Member State and vice versa, as well as within ports of call under the jurisdiction of a Member State. It would extend to all GHG emissions, including importantly methane, from ships of 5,000 GRT or above. The rapporteur noted that methane emissions had an enormous impact on climate change, since the GHG potential of methane was 87 compared to CO_2[29] on a 20-year time-frame; hence, she said, stricter regulation was needed.

Second, a maritime transport decarbonisation fund should be set up. This should foster research and development in the energy efficiency of ships; support investment in innovative technologies and infrastructure to decarbonise maritime transport, including short sea shipping and ports; and further the deployment of sustainable fuels. It should be established for the period from 2021 to 2030 and financed from the revenues of the ETS. In principle, shipping companies would be subject to the default ETS rules and thus have to purchase and surrender ETS allowances (known as EUAs). They would, however, by derogation from the default ETS rules, be entitled to pay an annual membership contribution to the Fund in accordance with their total emissions reported for the preceding calendar year,[30] the Fund in turn surrendering allowances collectively on behalf of member companies. The contribution per tonne of emissions should be set by the Fund by 28 February each year, and should be at least equal to the market price for allowances in the preceding year.

Third, there should be a target of reduction of CO_2 emissions from transport work amounting to at least 40% by 2030 compared with the first reporting year of the MRV (2018). This would be significantly tougher than the IMO target which took 2008, a year of low CO_2 emissions from ships due to the effects of the financial crisis.

Fourth, the Commission should set targets for member states for deployment of shore-side electricity, since the rapporteur saw huge potential in shore-side power and possible zero-emission ports. If ships at berth were required to switch off their engines and connect to the land grid, or use other energy sources with equivalent effect, this measure would, it was said, provide immediate health benefits to the citizens living in port areas.

The European Parliament has since taken up the rapporteur's ideas. On 16 September it voted in favour of a 40% reduction in CO_2 by 2030 for all maritime transport, and for the inclusion of ships of 5,000 GRT and over in the ETS; it also voted for the establishment of an "Ocean Fund" to run from 2022 to 2030 to contribute to protecting marine ecosystems. The Parliament's amendments will now be negotiated with the Council, under the Commission's direction and mediation, in a process of trilogue meetings.

One concern being raised in connection with this is the possibility of "carbon leakage", whereby ships might hypothetically attempt to reduce their ETS obligations by simply adding a stopover in a nearby non-EEA port. This would theoretically limit their ETS compliance

27 2003/96/EC.

28 See Art. 14(1)(c)

29 CO_2, by definition, has a GWP of 1 regardless of the time-frame used, because it is the gas being used as the reference.

30 Under Regulation (EU) 2015/757.

costs. The MRV defines "port of call" as a port where a ship stops "to load or unload cargo or to embark or disembark passengers". (Jutta Paulus' report, it should be noted, recommends the addition of the word "substantial" in relation to cargo.) Therefore, stops for all other purposes (e.g. refuelling, obtaining supplies, relieving the crew, or sheltering from adverse weather) cannot be used as pretext to avoid obligations under the EU maritime ETS. An evasive decision to make an artificial cargo stop outside the EU remains possible; but this would be unattractive to ship operators as its costs (extra fuel burn, additional port dues, opportunity costs, etc.) would, on average, be larger than the saving in ETS compliance costs. It is worth noting that in the build-up to the introduction of a Sulphur Cap in ECAs in 2015, there was much talk of possible modal shifts, but in the event these did not materialise.

As far as the UK is concerned, the relevant Brexit regulation[31] amends Part 2 of the Merchant Shipping (Monitoring, Reporting and Verification of Carbon Dioxide Emissions) and the Port State Control (Amendment) Regulations 2017[32] and transmutes the MRV into domestic law, with the intent of maintaining the EU MRV for vessels calling at UK ports following Brexit. Vessels visiting both EU ports and UK ports are thus now required to comply with both schemes. EU voyages and emissions continue to be reported to the European Commission while UK voyages and emissions have to be separately reported to the UK authorities. All other provisions of the MRV regulation continue to apply.[33]

In its Command Paper, *The Future Relationship with the EU: The UK's Approach to Negotiations*,[34] the government has stated that the UK would be open to considering a link between any UK ETS and the EU ETS (as Switzerland has done with its ETS), if it suits both sides' interests. If a linking agreement is not agreed, then either the UK ETS will operate as a standalone system, or a Carbon Emissions Tax will be implemented as an alternative means of ensuring a carbon price remains in place in all scenarios. There is no indication as to how either approach would treat emissions from international shipping. On 22 March 2021 the Minister for Aviation, Maritime and Security, Robert Courts, said the UK may follow the European Union's proposal to include shipping in its Emissions Trading System.[35]

10.6 UNCLOS – compatibility of the proposed EU measures

There is no problem under UNCLOS with the inclusion in the ETS system of voyages in and out of the EU, even though this will affect ships flagged in non-EU states, such as the UK. States have complete sovereignty over their internal waters, including ports, and this gives the port state jurisdiction over all vessels in those waters.[36] Regional measures, moreover, are permitted under Article 311(3).[37]

31 The Merchant Shipping (Monitoring, Reporting and Verification of Carbon Dioxide Emissions) (Amendment) (EU Exit) Regulations 2018, SI 2018/1388.

32 SI 2017/825.

33 The EU's proposed amendments to the MRV, which may lead to its modification with effect from 1 January 2022, come after the end of the Brexit transition period on 31 December 2021, and so will not affect the UK.

34 CP211, February 2020.

35 www.bloomberg.com/news/articles/2021-03-22/u-k-may-include-shipping-in-new-emissions-trading-platform

36 See UNCLOS, Arts. 2(1), 8, 11 and 12.

37 Art. 311(3) specifies that other agreements may modify or suspend provisions of the Convention, "provided that such agreements do not relate to a provision derogation from which is incompatible with the effective execution of the object and purpose of this Convention, and provided further that such agreements shall not affect the application of the basic principles embodied herein, and that the provisions of such agreements do not affect the enjoyment by other states parties of their rights or the performance of their obligations under this Convention".

Port sovereignty is admittedly subject to the publicity requirements in Art. 211(3) of UNCLOS as regards the prevention, reduction and control of pollution of the marine environment as a condition for the entry of foreign vessels into ports, and also to the principles preventing discrimination between vessels on the basis of nationality in Art. 227 and requiring good faith and non-abuse of rights in Art. 300. Inclusion of international shipping in the ETS on a non-discriminatory basis will not, however, violate these principles, even though it will have extra-territorial effects. This point has been well made in connection with another EU measure, namely the European Union's unilateral decision after the *Prestige* disaster to phase out single-hulled tankers, which occurred before the IMO regulated to this effect:

> The important question therefore is whether the EU double hull regulations adversely affect the object and purpose of the Convention or the rights of parties under the Convention. As presently drafted, the answer is almost certainly no. As we saw above, port states have jurisdiction under general international law to regulate matters affecting the good government of their ports – including pollution control and construction standards. UNCLOS confers no general right of entry to ports. Provided there is no discrimination in its application, a ban on single hull tankers in EU ports does not appear to violate UNCLOS. It thus entails no denial of UNCLOS rights and does not compromise the object and purpose of the Convention. But if any attempt is made to extend the EU regulation to enforce a ban on passage of single hull tankers in the territorial sea or EEZ, UNCLOS rights and the object and purpose of the Convention would be directly in issue, and then Article 311(3) could come into play.[38]

10.7 WTO – compatibility of the proposed EU measures

An issue that was raised when the IMO considered market-based mechanisms such as ETS arrangements was whether they would be subject to challenge as violations of WTO obligations to flag state members of that organisation. A fundamental principle of agreements made under the aegis of the WTO is the most favoured nation (MFN) principle. In particular, Art. 2 of the Agreement on Services (GATS) requires that any favourable treatment extended to any nation is also extended to all WTO members without discrimination, and that such treatment may exist either "in fact" or "in law". But it is hard to see how the EU measures could amount to discrimination. Where is the discrimination where a measure is applied in the same manner to all vessels or vessel owners regardless of their flag? In any event, there is a Decision of the Council of Trade in Services of 28 June 1996[39] which suspends the application of the MFN principle to the shipping sector, excluding any commitments already undertaken by WTO members. The EU has undertaken commitments only in the internal waterways transport subsector, with regard to rental services with operators and rental of vessels with crew; it has not decided to open the international maritime transport sector within the context of the GATS. Although it is unlikely that GATS will be considered to apply, in any case it provides[40] for similar exceptions for matters such as safety and human health as appear in Art.XX of the GATT.

The provisions of the GATT as regards goods might also come into play. A possible source of non-compliance concerns the provisions of Art XI.I. This reads:

> no prohibitions or restrictions other than duties, taxes and other charges . . . shall be instituted or maintained by any contracting party on the importation of any product . . . of any other contracting party.

---

38 See A. Boyle, "EU Unilateralism and the Law of the Sea" (2006) International Journal of Marine and Coastal Law 21: 15, where this quote appears.

39 See S/L/24.

40 In Art. XIV.

This is subject to an exception where, as provided under Art. XIII.1, the importation of "like" products from all third countries is similarly prohibited or restricted. This would clearly be the case with the possible EU measures.

There is also Art V, which requires that members enable freedom of transit for goods, vessels and other means of transport. Denial of access to ports will effectively limit the importation of the goods aboard the vessel, but this would not be the case with charges levied on the vessel. Any charges would have to be unreasonable and favour one nation other another to fall foul of this: again, this is not the case with the proposed extension of the MRV.

One of the side-agreements to the WTO may be more contentious with regard to the proposed measures. This is the Agreement on Technical Barriers to Trade (TBT). The TBT Agreement applies to technical regulations, standards and conformity assessment procedures. If a vessel can be regarded as a "product", then its inclusion within the ETS would constitute a technical regulation. Art. 2.1 of the TBT Agreement, which covers rules for "government and private regulatory systems", requires members imposing technical regulations to do so in a manner that infringes neither the MFN nor the national treatment rule. Annex 1.1 defines a technical regulation as a document "which lays down product characteristics or their related process and production methods . . . with which compliance is mandatory". However, it is difficult to see how the TBT can have any application as a ship can hardly be viewed as a "product". Shipping is a service; and although standards can include product and service standards, the TBT Agreement explicitly states in Annex 1 that it does not cover services.

10.8 Contractual issues

The inclusion of shipping in the ETS will create additional costs for shipowners and charterers. For voyage charters to and from the EU this will undoubtedly be reflected in freight rates; and with time charters allowing trading to this area, clauses apportioning the costs are likely to be developed, along the lines of BIMCO's USGTT clause.[41] However, for existing time charters which run into the ETS period, assuming that the EU is a permitted trading area, there is no mechanism for owners to recover these additional costs. The implied indemnity will not work, as these costs will be regarded as the natural costs of trading, as was the case in *The Dimitris L*.[42] There the time charterers' orders to proceed to the United States did not entitle the owners to be indemnified against the cost of US Gross Transportation Tax. Christopher Clarke J stated:

> As to clause 8, in a broad sense the tax was incurred in consequence of the order to proceed to the US. But owners are not entitled to recover from charterers under the implied indemnity the ordinary expenses and losses of trading or those of which the owner has accepted the risk . . . In the case of voyages to or from the USA USGTT is, as it seems to me, an ordinary expense of trading and not an unusual feature. In any event the risk of incurring or being liable for this very tax has been dealt with by clause 112 and, to the extent that the owners are not entitled to recover under that clause, the risk of having to bear the cost of the tax is one which they must be taken to have accepted.[43]

However, the current proposed amendment to the MRV may directly impact time charterers. The matter turns on the definition of a ship-owning "company" which is subjected to the scheme. The Commission originally defined it as "the shipowner or any other

41 A clause dealing with the US Gross Transportation Tax for those vessels visiting the US.
42 [2012] EWHC 2339; [2012] 2 Lloyd's Rep. 354.
43 [2012] EWHC 2339; [2012] 2 Lloyd's Rep. 354 at [55] (citations omitted).

organisation or person such as the manager or the bareboat charterer, which has assumed the responsibility for the operation of the ship from the shipowner, and has agreed to take over all the duties and responsibilities imposed by Regulation (EC) No 336/2006". The EU Parliament's proposed definition is, however, wider. It refers to "the shipowner or any other organisation or person such as the manager; *the time charterer* or the bareboat charterer, which has assumed the responsibility for the *commercial* operation of the ship *and is responsible for paying for fuel consumed by the ship*". The omission of the reference to the Regulation, and the words in italics, may be important, since these could place both the reporting obligations under the MRV and the financial costs of the ETS on the time charterer, who would have no means of recovery from the shipowner. This is something that needs to be borne in mind in future negotiations of time charters, assuming the EU Parliament's version of the revised MRV comes into effect on 1 January 2022.

The proposed amendment is also unclear as to how the alternative candidates for being the "company" under the MRV will be ascertained in any particular case, and this is also something that owners and time charterers will need to address over 2021. Shipowners will doubtless argue that time charterers should constitute the "company" under their liability for port dues. These are any charges the vessel must pay before a vessel leaves a port. However, payment under the ETS would be on an annual basis, and it is therefore doubtful whether it would constitute a port due. The wording of the Parliament's amendment has the time charterer as an alternative candidate for the "company" and omits the words proposed by the Commission "and has agreed to take over all the duties and responsibilities imposed by Regulation (EC) No 336/2006" which would require positive agreement by a party other than the shipowner to stand in the role of the "company", something that is understandable with a bareboat charterer. In the absence of agreement, presumably responsibility would fall on the shipowner, but this is something that should be clarified in the trilogue procedure.

10.9 Conclusion

Climate change is the most pressing issue of our times, with the prospect of the global carbon budget for a 1.5°C rise over pre-industrial levels becoming exhausted by the end of the decade. International shipping will have to change. The IMO's 2030 targets only appear capable of achievement through speed regulation and operational efficiency measures. Left to the market and the IMO alone, this will not happen. A governmental regulatory nudge is going to be required, although no individual state can force change on its own. However, a multinational organisation of states in the developed world may be able to do so.

This is where the EU Parliament's proposed amendment of the MRV comes in, under which as from 1 January 2022 voyages into and out of the EU will be included in the ETS. If the MRV is amended accordingly, the message to shipowners, and also to time charterers, will be a simple one: "It will cost you". Very few of them will have either the ability or the appetite to participate in the purchase of or trade in emissions credits. Instead they will have the option of paying an amount based on their annual emissions on voyages into and out of the EU into the proposed Maritime Development Fund. These contributions will be the "stick" for ensuring emissions reductions on such voyages, with the Maritime Development Fund being the "carrot" with the prospect of developing alternative 'green' maritime fuels – of which the hydrogen fuel cell currently seems the most promising – as well as enabling the electrification of ports in the EU with "green" energy to power ships at berth through "cold ironing".

The position of the UK at the moment is unclear. It will be interesting to observe what policy choices it makes in its new status in a post-Brexit world. It will continue with the existing MRV, amended to function as domestic law. It intends to set up its own ETS regime, but there is also the possibility of introducing a carbon tax. How it will react to the amended MRV over the next 12 months is unclear. The trade agreement reached with the EU just before Christmas 2020 does not constrain the UK's choices on whether to include international voyages into and out of the UK in its ETS. If it did so, it would likely only operate as regards voyages into and out of the UK from and to non-EU destinations, so as to avoid double charging with the EU regime. However, it is possible that the UK government may see a possible competitive advantage for its ports in encouraging ships to discharge in Southampton, say, rather than Antwerp, and then tranship their cargoes to an EU port. The matter remains unclear: though it is worth noting that the introduction of the EU sulphur caps in the ECA in 2015 led to speculation about modal shift which never transpired.

On the other hand, we must remember that at the end of 2021 the UK is due to host COP 26 in Glasgow. This sort of undermining of the EU's ETS for shipping will hardly play well there; nor will it sit comfortably with the UK's newly-declared enhanced climate change mitigation ambitions for 2030.

Shipowners and time charterers should now start addressing the revised MRV, which has a greater potential disruptive effect than the IMO's Sulphur Cap which heralded 2020 (along with news of a novel respiratory virus in Wuhan). It will be time to dust off "slow steaming" clauses, and to develop clauses allowing claims to apportion ETS costs between shipowners and time charterers. The position is complicated by the EU Parliament's inclusion of time charterers as an additional party liable for contribution under the revised MRV.

In 2017 Professor Kevin Anderson of the Tyndall Institute at the University of Manchester observed that so far climate change had most affected those who were the lowest emitters. That year climate change related floods in Bangladesh, displacing 600,000 people, caused hardly a ripple in the newsrooms of the developed world.[44] Wildfires in the US and Australia indicate that all this is beginning to change, and that by the end of the decade the impacts of global warming will be starting to pinch in Europe. A little reported item in the *Financial Times* for 17 August 2020 about the drying-up of rivers in France and Germany due to climate change noted:

> While summer thunderstorms have provided sporadic relief for parched fields in the past week, farmers, scientists and politicians say global warming is triggering multiyear droughts – 2018 and 2019 were also dry – and changing the climate of continental Europe in ways that will affect agriculture and the rest of the economy. This year's July was the driest in France since 1959 according to the national weather office, with less than a third of normal rainfall, while the average temperature between January and July was the highest since its records began."

It is time for international shipping to do its bit. The International Transport Forum's 2018 report "Decarbonising Maritime Transport Pathways to Zero-carbon Shipping by 2035"[45] shows how that can be done, outlining four viable pathways for substantial GHG reductions from the sector by 2035, but regulatory input is going to be needed to achieve this. The EU's revised MRV may well be just the start of a process to make sure that such an input does indeed eventuate.

44 www.newsecuritybeat.org/2017/08/flooding-bangladesh-calling-climate-change-high-ground/ (accessed March 2021).

45 "Decarbonising Maritime Transport: Pathways to Zero-carbon Shipping by 2035" at www.itf-oecd.org/decarbonising-maritime-transport-2035 (accessed March 2021).

Postscript

On 14 July 2021 the EU Commission issued a 581-page document, which included a proposed new Directive amending the 2003 ETS Directive.[46] This is considerably less extensive that the proposed amendment to the 2015 MRV Regulation, which the EU Parliament voted for in September 2020.

Maritime transport would then fall within the revised Directive (inserted Articles 3g to 3ge) which will apply in respect of emissions from intra-EU voyages; half of the emissions from extra-EU voyages; and emissions occurring at berth in an EU port. This rows back from the Parliament's earlier proposed amendments to the 2015 MRV Regulation which would have included all emissions from extra-EU voyages that started from or ended within the EU.

If the newly proposed revisions are ratified, the same rules that apply to other sectors covered by the EU ETS will then apply to maritime transport with regard to auctioning, the transfer, surrender and cancellation of allowances, penalties and registries (Article 16). Shipping will enjoy phased entry into the ETS. Shipping companies shall be liable to surrender allowances according to the following schedule: (a) 20% of verified emissions reported for 2023; (b) 45% of verified emissions reported for 2024; (c) 70% of verified emissions reported for 2025; (d) 100% of verified emissions reported for 2026 and each year thereafter. This is somewhat different from the inclusion in the ETS as of 1.1.2022 proposed by the EU Parliament. The current MRV Regulation applies only to CO2 emissions and the Commission leaves extension to other gases to a later phase, once the monitoring approaches and emissions factors of these gases has been agreed.

The proposed amending directive includes new definitions for "shipping company" and "administering authority in respect of shipping companies" in Article 3(v) and Article 3(w) respectively. The new definition covers the shipowner or any other organisation or person, such as the manager or the bareboat charterer, that has assumed the responsibility for the operation of the ship from the shipowner and that, on assuming such responsibility, has agreed to take over all the duties and responsibilities imposed by the International Management Code for the Safe Operation of Ships and for Pollution Prevention. However, the definition does not cover time charterers who would have become responsible under the Parliament's previously-proposed amendment to the MRV Regulation.

There have also been changes in the UK as regards international shipping and GHG reduction. On 20 April 2021 the Government announced that the Sixth Carbon Budget would include the UK's share of international shipping emissions, as recommended by the Climate Change Committee (CCC). The projections for international shipping emissions represent the estimated emissions from fuel sold in the UK for use in international shipping. However, shipping is not currently included in the UK's ETS.

46 Brussels, 14.7.2021,COM(2021) 551 final. 2021/0211 (COD). https://ec.europa.eu/info/sites/default/files/revision-eu-ets_with-annex_en_0.pdf<accessed 15.7.2021>.

CHAPTER 11

International legal aspects of Arctic shipping

*Youri van Logchem**

11.1 Introduction

Climate change is having a pronounced impact in the Arctic region. Average temperatures in winter are dropping, and the yearly seasonal variations in ice cover are increasing, which will result in less ice coverage.[1] Although there is no universally accepted definition of "the Arctic", its spatial scope is most often defined as the area around the North Pole.[2] At the heart of the Artic is the Arctic Ocean. Its warming has caused the ice cover to shrink, whereby waters that have historically been difficult to penetrate are now becoming more easily accessible.[3] The increased accessibility of the Arctic waters has functioned as a lure for conducting economic activities, including utilising these waters for (international) shipping purposes. As navigational uses rise, the historically grown perception of international shipping in the Arctic being an untenable prospect is increasingly challenged.[4] And if predictions come to fruition that the ice cover will continue to recede, this perception will incrementally only lose its force further, as the navigability of Arctic waters increases.[5] The driving factor behind these fundamental changes in the Arctic region is climatic change. In addition to its many negative consequences, new opportunities are also created by climate change, including that it facilitates the creation of potentially new international shipping routes that were previously inaccessible, or that it makes those already in existence better accessible. At the same time, climate change creates new threats, or exacerbates already existing ones in the Arctic context, including drifting hard multi-year ice.

But what is the rationale underlying Arctic shipping? The key to this answer lies in economic reasons. Because of its strategic placement, the Arctic Ocean offers shorter shipping routes between the Pacific Ocean and the Atlantic Ocean. Their use would translate in reduced fuel consumption and thus lowering operation costs for ships. A figure that is often

---

* Member of the Institute of International Shipping and Trade Law, Swansea University. The author would like to thank Jessica Schechinger for her invaluable comments. Any faults or omissions are, of course, the sole responsibility of the author.

1 JE Kay et al., "Inter-annual to Multi-decadal Arctic Sea Ice Extent Trends in a Warming World" (2011) 38 GRL L15708.

2 EJ Molenaar, "Arctic Marine Shipping: Overview of the International Legal Framework, Gaps, and Options" (2009) 18(2) JTLP 291.

3 P Barkham, "Russian Tanker Sails Through Arctic without Icebreaker for First Time" *The Guardian* (London, 24 August 2017) at www.theguardian.com/environment/2017/aug/24/russian-tanker-sails-arctic-without-icebreaker-first-time (accessed 2 June 2021).

4 DR Rothwell, "International Law and Arctic Shipping" (2013) 22(1) MSILR 67, 67–68.

5 TC Stevenson et al., "An Examination of Trans-Arctic Vessel Routing in the Central Arctic Ocean" (2019) 100 MP 83; M Byers and E Lodge, "China and the Northwest Passage" (2019) 18(1) CJIL 57, 58.

 DOI: 10.4324/9781003155195-11

invoked to highlight the potential benefits of Arctic shipping is the following: traversing through the Northwest Passage (NWP), instead of travelling via the Panama Canal or along the Cape Horn route, cuts a ship's journey by approximately a third. To be more specific, and depending on the route, the average shipping distances will be reduced by respectively 9,000 or 17,000 kilometres.[6] As a further benefit, the NWP would be able to facilitate the voyages of larger ships than the Panama Canal.[7]

Furthermore, the increased prospect of Arctic shipping can be considered both a boon and a bane. On the one hand, there are the economic benefits that it may provide for both shipping companies and Arctic coastal States, with the latter being able to attract more trade to their waters and ports. But, on the other hand, shipping threatens the Arctic marine environment.[8] The latter is relatively pristine and highly vulnerable to the effects of economic activities.[9] An oil spill, as the breaking down of the Exxon Valdez in 1989 vividly illustrates, would have disastrous results for the Arctic environment and its ecosystems, as oil and other pollutants degrade much slower in an ice-covered area. Recognising the sensitive nature of the Arctic maritime environment, some of its coastal States have resorted to prescribing environmental standards for ships operating in Arctic waters that extend beyond the generally accepted international rules and standards regarding vessel-source pollution. This creates a tension with the navigational rights and freedoms that States have under international law.[10] Closely tied to the extent to which coastal States are permitted to adopt unilateral measures to the aim of protecting the marine environment, is that this varies with the maritime zone involved. And here lies a further difficulty with regard to parts of the Arctic region, in that the status of some of its waters, including the NWP and the Northern Sea Route (NSR), has been the subject of different views. Particularly controversial has been the designation by, respectively, Canada and the Russian Federation of both the NWP and parts of the NSR as being part of their internal waters. At the root of these assertions lie either a claimed historic title over these waters, or that the States concerned could draw straight baselines along their Arctic coasts.[11] Its controversial nature can be inferred from the fact that pursuant to these designations, the waters in question have been placed under the complete control of Canada and the Russian Federation, whereby navigational rights and freedoms for other States are severely impacted; that is, in a way that none may exist for them under international law.

The discussions that arose around these two issues, of whether it is lawful under international law for coastal States to extend their domestic legislation regarding vessel-source pollution beyond the generally accepted international rules and standards, and what the status is of the NWP and NSR, continue to the present day. These debates are

6 KJ Wilson et al., "Shipping in the Canadian Arctic: Other Possible Climate Change Scenarios" (2004) IGARSS 1853, 1853–1854.

7 H Tuerk, "The Arctic and the Modern Law of the Sea" in JM van Dyke et al. (eds), *Governing Ocean Resources: New Challenges and Emerging Regimes* (Brill Nijhoff 2013) 118.

8 C Schofield et al., "Boundaries, Biodiversity, Resources and Increasing Maritime Activities: Emerging Oceans Governance Challenges for Canada in the Arctic Ocean" (2009) 34 VLR 43.

9 J Bai, "The IMO Polar Code: The Emerging Rules of Arctic Shipping Governance" 30(4) IJMCL 674, 676.

10 K Bartenstein, "Between the Polar Code and Article 234: The Balance in Canada's Arctic Shipping Safety and Pollution Prevention Regulations" (2019) 50(4) ODIL 335, 336; T Potts and C Schofield, "Climate change and evolving regional ocean governance in the Arctic" in H Scheiber and J Paik (eds), *Regions, Institutions and The Law of the Sea: Studies in Ocean Governance* (Brill Nijhoff 2013) 437, 453.

11 See section 11.4.1.

notoriously difficult and are therefore unlikely to be laid to rest any time soon. This is illustrated by their long lineage, which can be retraced to the end of the twentieth century, but also by the fact that either side to these debates can invoke sensible arguments in support of its position.[12] Therefore, this chapter does not aim to break new ground, also considering that the existing literature on these subjects is rather voluminous. Against this background, the chapter's aim is much more modest in that it provides a *tour d'horizon* of the various issues that arise from the perspective of international law (of the sea) in connection with shipping in the Arctic.

11.2 Roadmap to the chapter

The chapter is organised as follows. To begin with, the scene will be set, by sketching the general background to the issue of Arctic shipping and the international legal issues that it gives rise to. It will also lay the groundwork by giving an overview of the different shipping routes that exist in the Arctic Ocean, their historic and potential usage, and the specific legal questions that have surrounded them. Then, once the stage has been set, attention will be directed to the international legal framework applicable to Arctic waters. This legal framework consists of several layers. At the heart of the legal regime is the international law of the sea, which holds significant relevance for the Arctic Ocean. Its relevance is illustrated by that the law of the sea addresses navigational activities, but also because of its giving coastal States certain powers to combat vessel-source pollution. The second layer of legal rules that States that operate in the Arctic Ocean have to abide by are other international shipping regulations, including the various conventions that have been adopted under the auspices of the International Maritime Organisation (IMO) and which are commonly concerned with the safety of shipping.[13] A more recent addition to these shipping regulations is the Polar Code,[14] which involves amendments of two IMO Conventions: the 1973 Convention for the Prevention of Pollution of Ships (MARPOL)[15] and the International Convention for the Safety of Life at Sea Convention (SOLAS).[16] Domestic legislation adopted by States on the basis of Article 234 of the United Nations Convention on the Law of the Sea,[17] which is specifically tailored to ice-covered areas and enables the coastal States to adopt laws and regulation relating to marine pollution caused by vessels in the exclusive economic zone (EEZ), forms the

12 M Krafft, "The Northwest Passage: Analysis of the Legal Status and Implications of its Potential Use" (2009) 40(4) JMLC 537, 567.

13 A Chircop, "The Growth of International Shipping in the Arctic: Is a Regulatory Review Timely?" (2009) 24(2) IJMCL 355, 361.

14 Polar Code. (2014/2015). International Code for Ships Operating in Polar Waters (Polar Code), IMO Resolution MSC.385(94) (21 November 2014, entered into force 1 January 2017); Amendments to the International Convention for the Safety of Life at Sea 1974, IMO Resolution MSC.386(94) (21 November 2014, entered into force January 2017); Amendments to MARPOL Annexes I, II, IV and V, IMO Resolution MEPC.265(68) (15 May 2015, entered into force 1 January 2017) At www.icetra.is/media/english/POLAR-CODE-TEXT-AS-ADOPTED.pdf (accessed 9 June 2021).

15 International Convention for the Prevention of Pollution from Ships (adopted 2 November 1973/17 February 1978, entered into force 2 October 1983) 1340 UNTS 61 (MARPOL Convention).

16 International Convention for the Safety of Life at Sea (adopted 1 November 1974, entered into force 25 May 1980) 1184 UNTS (SOLAS).

17 United Nations Convention on the Law of the Sea (adopted 10 December 1982, entered in force 16 November 1994) 1833 UNTS (LOSC, or Convention).

third layer of rules that has to be addressed. The chapter concludes by critically reviewing the main elements of the legal framework dealing with Arctic shipping. And it ends with some remarks on whether the current framework is suitable to deal with the idiosyncrasies of the Arctic environment.

11.3 Setting the scene

Several shipping routes can be identified within the Arctic Ocean that could prove to be useful connections between different continents: that is, between Asia and Europe, the western part of North America to Europe, and the eastern part of North America to Asia. More specifically, within the Arctic Ocean there are two already existing routes (i.e., the NWP and the NSR), which are already navigable during summer and fall, if the weather allows, and one that looms on the horizon, if the effects of climate change on the Arctic region do not abate (i.e., the Central Arctic Ocean Route (CAOR)). Extreme weather conditions continue to render the use of the NWP and NSR intermittent, however. These routes have in common that they are almost exclusively located off the coast of one coastal State, being Canada and the Russian Federation. However, as aforementioned, there are differing views over the status of the NWP and parts of the NSR, and thus the legal regime that would be applicable therein.

At present, and despite that the ice cover is undoubtedly thinning, the heart of the Arctic Ocean, where the future CAOR would come to lie, remains largely impenetrable due to present ice cover. However, it has been predicted that this is going to change. If the warming of the Arctic region continues at its current pace, a CAOR could already emerge within the next several decades.[18] More specifically, predictions, although showing some variations, indicate that this could happen as early as 2040.[19] Others have predicted that a route taking vessels through the centre of the Arctic Ocean, unescorted by icebreakers, might emerge no sooner than 2050.[20] Once the CAOR emerges, it does come with a caveat and that is its overall navigability. This will remain highly variable, due to ice conditions that will continue to differ annually, as well as because of the dangers posed by drifting hard multi-year ice.

To put the usage of the Arctic waters for shipping purposes first into a broader perspective. On the whole, the annual window for navigation in Arctic waters seems to be increasing, whereby a further intensification of shipping traffic in the region can be assumed. Yearly variations in ice thickness will impact the overall reliability and navigability of the Arctic shipping routes for some time to come, however. Further questions remain around whether Arctic shipping is going to take actual flight, particularly in the short term;[21] this is despite some firmer claims to the contrary.[22] Generally, the main variable tied to whether

18 N Melia et al. (2016), "Sea Ice Decline and 21st Century Trans-Arctic Shipping Routes" 43 GRL 9720–9728.

19 J Stroeve and D Notz, "Changing State of Arctic Sea Ice Across All Seasons" 13(1) ERL 103001.

20 D Notz and SIMIP Community, "Arctic Sea Ice in CMIP6" (2020) 47 GRL 1–11.

21 KE Skodvin, "Arctic Shipping – Still Icy" in MH Nordquist et al. (eds), *Challenges of the Changing Arctic: Continental Shelf, Navigation, and Fisheries* (Brill Nijhoff 2016) 150; Potts and Schofield, see note 10, pp. 452–454.

22 See e.g., E Bekkers et al., "Melting Ice Caps and the Economic Impact of Opening the Northern Sea Route" (2018) 128 TEJ 1095, 1096.

Arctic shipping will become commonplace seems to be the ice cover, and the pace at which it is going to retract. Due to there being a host of variables involved, this is difficult to predict with any precision.[23] Other general features of the Arctic shipping routes are their interannual variability and that a general uncertainty surrounds their accessibility, both of which can be linked to the prevalent ice conditions as well.[24]

Shrinking ice cover does not automatically lead to the creation of a shipping environment that is devoid of risks. Rather the contrary, as the ice caps continue to melt, a new serious threat to shipping is formed. Once the softer first year ice melts, multi-year ice will start to break off, and drift through the Arctic Ocean. Drifting multi-year ice poses a serious hazard to shipping, one that climate change will only exacerbate. Drift ice will increase the risk of collisions, and will impede navigation by blocking Arctic passages.[25] The latter happened in 2007, for example, which was a year when ice conditions were seemingly favourable for navigation in the Arctic Ocean. However, pieces of hard multi-year ice that had broken off started to drift through the Arctic Ocean and ultimately filled up the NSR, blocking its navigational routes in the process.[26] Furthermore, the inherent risks that flow from navigating in harsh and rather unpleasant weather conditions will not disappear either, as the Arctic waters will continue to provide a quite inhospitable environment for shipping activities.[27] Two illustrations of that ship operations in Arctic waters can be considered riskier are as follows: first, the adverse weather conditions that ships have to cope with; and, second, the fact that map coverage of these waters is incomplete. Available rescue services in the region are also more limited. Although technological advancements have alleviated some of the inherent risks associated with Arctic navigation, such technology does come at a greater cost.[28]

Other factors that reduce the benefits flowing from a reduction of the amount of fuel and time that a voyage through Arctic waters requires are, for example, insurance premiums, the costs of which are higher for ship operators that travel through one of the Arctic shipping routes than they would be elsewhere. Other costs that have to be borne by the ship operators to enable Arctic shipping are, for example, hiring icebreakers, whose assistance is usually required to successfully complete a transit, or the payment of transit fees.[29] All these aspects may counteract some of the advantages associated with Arctic shipping. It could also be that Arctic shipping might prove beneficial for a more limited range of types of shipping. For example, vessels carrying their cargo in bulk might benefit from utilising

23 DL VanderZwaag, "Climate Change and the Shifting International Law and Policy Seascape for Arctic Shipping" in RS Abate (ed), *Climate Change Impacts on Ocean and Coastal Law: U.S. and International Perspectives* (OUP 2015) 301; J Kraska, "Governance of Ice-Covered Areas: Rule Construction in the Arctic Ocean" (2014) 45 ODIL 260.

24 A Chircop, "Climate Change and the Prospect of Increased Navigation in the Arctic" (2007) 6(2) WMU JMA 193, 196–198.

25 V Gavrilov et al., "Article 234 of the 1982 United Nations Convention on the law of the Sea and Reduction of Ice Cover in the Arctic Ocean" 106 MP 1, 5.

26 N Melia et al., see note 18, p. 9720.

27 Skodvin, see note 21, p. 150; L Zou, "Comparison of Arctic Navigation Administration between Russia and Canada" in MH Nordquist et al. (eds), *Challenges of the Changing Arctic: Continental Shelf, Navigation, and Fisheries* (Brill Nijhoff 2016) 288–289.

28 Potts and Schofield, see note 10, p. 258; J Kraska, "The Law of the Sea Convention and the Northwest Passage" (2007) 22(2) IJMCL 257, 258.

29 J Saul, "Arctic Headache for Ship Insurers as Routes Open Up" *Reuters* (London, 27 October 2020) at www.reuters.com/article/us-climate-change-arctic-shipping-insura-idUKKBN27C1P1 (accessed 3 June 2021).

the Arctic shipping routes, whereas vessels carrying containers, but whose usual *modus operandi* is not based on travelling between two points, may incur higher costs.[30] A further variable as to whether Arctic shipping would be profitable is that this will vary according to the port destination involved. To give an example, if a vessel leaving a North American port has a more southerly located port as its destination, for instance the port of Shanghai (China), then a journey through Arctic waters would slightly prolong the journey compared to the route leading ships through the Suez Canal.[31]

At the same time, ship traffic within the Arctic region raises various concerns.[32] One of these centres around the Arctic's marine environment and its ecosystems, which remain relatively pristine. Its vulnerability to the effects of economic activities, including shipping, has been well documented and has created a more general anxiety on the part of Arctic States and their populations around the occurrence of such activities.[33] A particular concern connected to shipping is pollution emanating from vessels, which may overwhelm the adaptive capacities of the Arctic ecosystems, and reduce, or even eliminate entire populations of living resources. Carbon emissions and alien animal species that could be introduced through vessel operations into the ecosystems are further threats linked to Arctic shipping.[34] There is thus a clear environmental imperative to protecting the Arctic marine environment from the adverse effects of shipping. Arguably, this imperative transcends the exclusive interests of the Arctic coastal States.[35]

None of the Arctic coastal States have prohibited navigation generally. However, two of the Arctic States (i.e., Canada and the Russian Federation), fully aware of the threats that shipping poses, have adopted domestic environmental legislation with the aim of ensuring safe Arctic navigation that have no equivalent on the international level, which has raised questions around whether some of these domestic approaches can be reconciled with international law. For example, in Arctic waters located off the Canadian archipelago, Canada introduced a mandatory reporting system for vessels weighing more than 300 tonnes when operating within a distance of 200 nm from its coast through its Northern Canada Vessel Traffic Services zone (NORDREG).[36] Several motivations lie behind the NORDREG, all of which are somehow connected to the protection of the Arctic marine environment. It has been created to increase the safety of navigation in the region, improve incident responsiveness, and it enables Canada to keep a close eye on vessels carrying noxious or dangerous cargos that, in the event of an incident, would severely pollute the Arctic marine environment. The responsibility for enforcement of the NORDREG lies with the Canadian coast guard. Canada has invoked Article 234 of the LOSC as the legal basis underpinning the NORDREG, which means that Canada did not deem it necessary to involve the IMO prior to its adoption.[37] Since its introduction, the NORDREG has attracted criticism from

30 Skodvin, see note 21, p. 151.

31 N Melia et al., see note 18, p. 9726.

32 Kraska, see note 28, p. 258.

33 Bai, see note 9, p. 676.

34 Arctic Council, *Arctic Marine Shipping Assessment 2009 Report* (PAME) 8–9 At www.pame.is/images/03_Projects/AMSA/AMSA_2009_report/AMSA_2009_Report_2nd_print.pdf (accessed 5 June 2021).

35 Bai, see note 9, p. 687.

36 Northern Canada Vessel Traffic Services Zone Regulations, SOR/2010–127, 1 July 2020 at https://laws-lois.justice.gc.ca/PDF/SOR-2010-127.pdf (accessed 5 June 2021); J Kraska, "The Northern Canada Vessel Traffic Services Zone Regulations (NORDREG) and the Law of the Sea" (2015) 30(2) IJMCL 225, 227–228.

37 Ibid., pp. 237–238.

various sides, however, including the Baltic and International Maritime Council (BIMCO), Singapore,[38] the United States,[39] as being problematic from the view of international law.[40]

11.3.1 Northwest passage

The NWP starts from the Pacific Ocean, and after weaving through the Canadian archipelago via a complex network of passages, it extends into the Atlantic Ocean. While coastal shipping in the NWP is a more common occurrence, it is only more seldom used for international shipping activities. Navigation within the NWP does not occur via a fixed route.[41] Rather, it is home to at least seven routes that could be utilised for shipping purposes.[42] The main variables involved, as to whether an *individual route* within the NWP is navigable, are the weather and prevailing ice conditions.

In September 2013, the NWP was first transited by a considerably sized commercial vessel. Then, the *Nordic Orion* carrying a cargo of coal it had loaded in Vancouver (Canada) successfully navigated the NWP to deliver its load of coal to Pori (Finland).[43] It was subsequently hailed as a sign of more things to come.[44] However, there is a twist to this story that does not neatly fit into this narrative of that a sudden increase in shipping activity within the NWP is to be expected. It was only because the Canadian coast guard offered its icebreaking services for free that the Danish company owning the *Nordic Orion* (i.e., Nordic Bulk Carriers) had decided to include the NWP in its route. Had this been different, the costs of the voyage would have been simply too high to make it worthwhile to transit through the NWP.[45] The journey of the *Nordic Orion* was carried out in accordance with Canada's 1970 Arctic Waters Pollution Prevention Act (AWPPA), on which more follows.[46]

There are different views around the status of the NWP, which fall along the following two lines. Canada's position regarding the NWP is that it is part of its internal waters, to which end it has invoked two separate legal bases. First, the existence of a historic title over these waters; and, second, that Canada is allowed to establish straight baselines around its Arctic Archipelago, whereby the waters lying landwards thereof are transformed into internal waters.[47] On the other side of the debate there is the position, held by the United

38 Singapore, Statement to MSC, in Annex 28 of IMO, Report of the Maritime Safety Committee in its Eighty-Eighth Session, IMP Doc. MSC/87/26 (15 Dec. 2010), [11.36].

39 US State Department, Digest of United States Practice in International Law, 2010(2010), 518, at https://2009-2017.state.gov/documents/organization/179316.pdf (accessed 9 June 2021).

40 Tuerk, see note 7, p. 126.

41 Chircop, see note 24, p. 195; DR Rothwell, "The United States and Arctic Straits: The Northwest Passage and the Bering Strait" in D Pharand et al. (eds), *International Law and Politics of the Arctic Ocean: Essays in Honor of Donat Pharand* (Brill Nijhoff 2015) 166.

42 D Pharand, "The Arctic Waters and the Northwest Passage: A Final Revisit" (2007) 38(1–2) ODIL 3, 29–30.

43 F Lasserre and O Alexeeva, "Analysis of Maritime Transit Trends in the Arctic Passages" in D Pharand et al. (eds), *International Law and Politics of the Arctic Ocean: Essays in Honor of Donat Pharand* (Brill Nijhoff 2015) 184.

44 "Historic Sea Route Opens Through Canadian Arctic Waters" (The Maritime Executive, 25 September 2013) at www.maritime-executive.com/article/Historic-Sea-Route-Opens-Through-Canadian-Arctic-Waters-2013-09-25 (accessed 5 June 2021).

45 Lasserre and Alexeeva, see note 43, 187.

46 "Historic Sea Route Opens Through Canadian Arctic Waters", see note 44.

47 Canada's Secretary of State for External Affairs, Joe Clark, Statement on Sovereignty, Hansard, 33rd Parliament, Vol. 5 (10 Sept. 1985), 6463 at http://parl.canadiana.ca/view/oop.debates_HOC3301 (accessed 5 June 2021).

States, that the NWP is an international strait in the sense of Articles 37 and 38(2) of the LOSC, which means that an international transit passage regime would apply there.[48]

For much of the twentieth century, the status of the NWP was an issue that failed to gather much attention.[49] However, this changed on 5 September 1969 when the NWP was transited by a commercial merchant ship for the first time. Then, the *SS Manhattan*, a reinforced super-tanker that was flagged to the US, made a successful journey through the NWP. During its voyage, it was accompanied by two Canadian icebreakers, a vessel of the Canadian coast guard (*J.A. Macdonald*), and a Canadian observer was furthermore present on board the *SS Manhattan*. The main purpose of the *SS Manhattan* was to gauge whether the route could be used for transporting oil, but it was ultimately considered commercially unviable due to the high(er) costs involved. Spurred on by public opinion, which woke up to the possibility that the use of the NWP by ships would intensify and of the potential detrimental impact that this could exert on the pristine Arctic marine environment – and in light of that a second voyage of the *SS Manhattan* was upcoming – Canada was forced into formulating a response.[50] This response came in the shape of the 1970 AWPPA.[51] Introduced in 1970, and thus prior to that negotiations at the Third Conference on the Law of the Sea (1973–1982) (UNCLOS III) had begun, strict environmental and construction standards were created for private vessels operating within 100 nautical miles (nm) off the Canadian coast. In 2009, the scope of the AWPPA was revised to include Canada's Arctic waters up to the 200 nm EEZ limit.[52] Ever since its enactment, the AWPPA has been surrounded by controversy, as several States protested.[53] Among these States, the United States has been particularly vocal in its objections. Its main concern centres around the impact of the AWPPA on its navigational freedoms within Arctic waters. According to the United States, Canada's unilateral assertion of having jurisdiction over what was in fact the high seas, and where there is freedom of navigation, was contrary to international law.[54]

After the voyage of the *SS Manhattan*, which had raised the possibility of the NWP becoming a navigational hub in the Arctic Ocean, a period of relative calm followed. This calm came to an end in 1985, when the debate over the status of the NWP flared up once again between Canada and the United States. Then, the United States Coast Guard icebreaker *Polar Sea* sought to navigate through one of the routes in the NWP. Prior to its transit, the United States gave notification thereof to Canada. In a demarche, the United States observed that there was no notification requirement for this voyage under international law;[55] although it recognised that Canada's position could well be at variance therewith. However, the United States indicated that it would be open to Canadian officials being placed on board during the transit, or Canada's participation in the transit more generally. None of

48 Office of Ocean Affairs, Limits in the Seas No. 112, United States Responses to Excessive National Maritime Claims (1992), 29–30 at www.state.gov/wp-content/uploads/2019/12/LIS-112.pdf (accessed 5 June 2021).

49 Rothwell, see note 41, pp. 166–167.

50 A de Mestral, "Article 234 of the United Nations Convention on the Law of the Sea: Its Origins and Its Future" in D Pharand et al. (eds), *International Law and Politics of the Arctic Ocean: Essays in Honor of Donat Pharand* (Brill Nijhoff 2015) 111; Rothwell, see note 41, p. 167.

51 Canada, Arctic Waters Pollution Prevention Act, Statutes of Canada 1969–1970, chapter 47.

52 Bartenstein, see note 10, pp. 335–336.

53 De Mestral, see note 50, p. 112.

54 Pharand, see note 42, p. 12; A Roach and RW Smith, *Excessive Maritime Claims* (3rd ed, Martinus Nijhoff 2012) 319.

55 M Byers, *International Law and the Arctic* (CUP 2013) 134–136.

these options was deemed detrimental by the United States to its legal position concerning the NWP and its claim that a transit passage right exists therein. Not only was this offer not acceptable to Canada, but the latter's position turned out to be more broadly at odds with the position of the United States. Canada indicated that the mere giving of notification was insufficient. Central to its position was the view that the waters concerned were part of Canada's historic internal waters, by which token obtaining prior authorisation was necessary. However, the United States refused to ask for such prior authorisation for reasons of it considering the NWP to be an international strait under Articles 37 and 38(2) of the LOSC, where a transit passage right exists for vessels flying the United States flag. Since then, the United States has reiterated its position on the NWP as being an international strait on several occasions.[56] The success of the United States' position is tied to the NWP satisfying the functionality requirement, which is one of the necessary attributes a strait must possess in order to qualify as an international strait for the purposes of Articles 37 and 38(2) of the LOSC (see section 11.4.4), and herein lies a potential rub. Historically, international passages through the NWP have been more limited, although, amongst others, government vessels of the United States, particularly its submarines, have utilised the route more frequently.

To now return to the *Polar Sea* and its held intention to transit through the NWP. In the end, the voyage went ahead, as Canada and the United States settled on allowing it to take place without prejudice to their respective legal positions on the status of the NWP. In its wake, Canada took several additional controversial steps, however, whereby it sought to placate public opinion, which had been inflamed by the event, and to broadcast the message to the international community at large of the Arctic waters being under its sovereignty.[57] One of the concrete steps that Canada took was to draw straight baselines around those parts of the Canadian coast that abut on the Arctic Ocean, as a consequence of which the waters located landwards of the baselines became internal waters (see section 11.4.1). In 1988 Canada and the United States agreed to the 1988 Agreement on Arctic Cooperation (1988 Agreement).[58] Its conclusion must be read against the backdrop of the *Polar Sea*'s voyage, and the clearly felt need by Canada to formulate a response to that event. In a nutshell, the 1988 Agreement is concerned with the conducting of marine scientific research. Under the agreement, the United States agreed to request prior consent from Canada if the former's icebreakers intended to navigate through sea areas which Canada perceives as its internal waters.[59] However, the broader context of the 1988 Agreement illustrates that this does not involve an alteration of the United States' position that there is right of transit passage within the NWP.[60]

In only one of the aforementioned voyages, that of the *Polar Sea*, no prior authorisation was sought by the United States from Canada. While this is true, a significant aspect is that during the transit of the *Polar Sea*, it was not only accompanied by Canadian icebreakers, but a Canadian observer was present on-board, which raises the suggestion that Canada implicitly agreed to the crossing.[61] However that may be, since then, the dispute between

56 Roach and Smith, see note 54, pp. 478–479.

57 Rothwell, see note 41, p. 169.

58 Agreement between Canada and the United States on Arctic Cooperation (11 January 1988) 1852 UNTS 59 (1988 Agreement).

59 Ibid., Article 3.

60 Tuerk, see note 7, p. 124; Pharand, see note 42, p. 12; Kraska see note 28, p. 258.

61 Pharand, see note 42, p. 39.

Canada and the United States over the status of the NWP has, once again, been largely dormant, as they have settled on a position of agree-to-disagree.[62]

But are changes afoot? Perhaps, given that as of more late the Arctic Ocean has seemingly risen higher on the political agenda of the United States.[63] To put matters into perspective, however, it is highly unlikely that the issue of the status of the NWP would devolve into a very significant bone of contention between the two States in the light of their historically cordial relations. However that may be, the *2013 National Strategy for the Arctic Region* shines particular light on what the position of the United States concerning the Arctic Ocean is, making it clear that it is built on there being the freedom of navigation for its vessels.[64] Also, there are other signs that the broader dynamics of the dispute as to the status of NWP are changing. Having been historically a largely bilateral affair between Canada and the United States, the issue around the status of the NWP might take on a multilateral character. Against the backdrop of climate change and the receding ice cover, other States have started to express interest in utilising the NWP as a shipping route. In this vein, China envisages using the NWP for international shipping purposes, and it clearly believes this to be a possibility that looms close on the horizon.[65] However, so far, it has kept its legal position as to the status of the NWP vague.[66] The Russian Federation has not contradicted the validity of Canada's claim,[67] but its "support" for the latter's position is not surprising, however, considering its role in the negotiation process of Article 234 of the LOSC and that both States have created comparable regimes in relation to the Arctic waters lying off their coasts.[68] A common aspect of their introduced legal regimes is that the navigational rights and freedoms of third States are reduced for environmental reasons.

11.3.2 Northern Sea Route

There is a clear correlation in the NSR between receding ice cover and rising levels of shipping activity. Comparatively, it has witnessed a higher volume of international commercial shipping than the NWP. This is a trend that is likely to continue, as ice conditions in the NSR are generally less severe compared to those in the NWP.[69] Therefore, in the short run, the NSR will hold greater appeal for international shipping purposes.[70] One of the main benefits of using the NSR for inter-Arctic shipping lies in that it may provide a significantly faster alternative to the Suez Canal route.[71] At present, and to place shipping in the NSR in perspective, mainly Russian flagged vessels, or foreign flagged vessels having a Russian

62 Bartenstein, see note 10, p. 347; Kraska, see note 28, pp. 268–269.

63 United States, "Arctic Region Policy", National Security Presidential Directive/NSPD – 66; Homeland Security Presidential Directive/HSPD – 25 (9 January 2009) at www.fas.org/irp/offdocs/nspd/nspd-66.htm (accessed 5 June 2021).

64 President of the United States, "National Strategy for the Arctic Region" (May 2013) at https://obamawhitehouse.archives.gov/sites/default/files/docs/nat_arctic_strategy.pdf (accessed 5 June 2021).

65 Byers and Lodge, see note 5, p. 69.

66 Byers, see note 55, p. 139.

67 Byers and Lodge, see note 5, p. 60.

68 P Fields, "Article 234 of the United Nations Convention on the Laws of the Sea: The Overlooked Linchpin for Achieving Safety and Security in the U.S. Arctic?" (2016) 7(1) HNSJ 55, 75.

69 Pharand, see note 42, p. 3.

70 Chircop, see note 24, p. 195.

71 Bekkers et al., see note 22, p. 1095; Zou, see note 27, p. 287.

port as their final destination, include the NSR in their shipping route.[72] Inter-ocean transit by foreign flagged vessels remains more infrequent, however.

The year 2009 was hailed as the year in which the NSR was first successfully used for a foreign commercial inter-ocean transit.[73] This is when two commercial vessels (the *Beluga Fraternity* and *Beluga Foresight*), both of relatively small size and which belonged to Germany's Beluga Shipping, successfully navigated the NSR.[74] They collected their cargo in South Korea, to subsequently make their way to the Russian port of Vladivostok. There, the two vessels were joined by Russian icebreakers, after which they set sail for the port of Rotterdam, the Netherlands.[75] A year later, in 2010, a larger bulk carrier (the *Baltica*) transited the NSR successfully.[76] In the following year the *Vladimir Tikhonov*, a tanker of over 163,000 deadweight tonnes, took on the title of the largest vessel having successfully navigated the NSR.[77]

Shipping within the NSR is variable throughout the year, as available routes change according to the weather and ice conditions.[78] Whereas in severe ice conditions, the route lies near the Russian Federation's coastline, when weather conditions are more benign, the route could be within its EEZ or even in the high seas. Seasonal changes in NSR route availability do not change the applicable legal regime, however. Navigation in the NSR *simpliciter* has been placed by the Russian Federation under its control;[79] this is thus irrespective of whether it factually occurs in its internal waters, the territorial sea, in the EEZ, or on the high seas.[80] In fact, route variability is one of the motivating reasons for the Russian Federation taking this sweeping approach to navigation in the NSR.[81]

Three possible legal justifications have been identified in legal literature that underpin the position of the Russian Federation as to why it is allowed under international law to comprehensively control navigation in the NSR.[82] First, that historically the NSR has been part of the Russian Federation's internal waters, meaning it has a historic title over them. This is illustrated by the fact that the Russian Federation has exercised full control over navigation in the NSR over a long period of time without protest from other States.[83] Second, as an Arctic coastal State, the Russian Federation carries special responsibilities for the Arctic waters and its ecosystems, by which token it must enjoy extensive

72 Lasserre and Alexeeva, see note 43, pp. 186–187.

73 Potts and Schofield, see note 10, p. 452; E Franckx and L Boone, "New Developments in the Arctic: Protecting the Marine Environment from Increased Shipping" in MH Nordquist et al. (eds), *The Law of the Sea Convention: US Accession and Globalization* (Martinus Nijhoff 2012) 188–190.

74 Potts and Schofield, see note 10, p. 452.

75 S Phalnikar, "German Commercial Ships Make Historic Arctic journey" Reuters (London, 12 September 2009) at www.dw.com/en/german-commercial-ships-make-historic-arctic-journey/a-4679955 (accessed 5 June 2021).

76 Potts and Schofield, see note 10, p. 452.

77 Lasserre and Alexeeva, see note 43, p. 187.

78 V Gavrilov, "Legal Status of the Northern Sea Route and Legislation of the Russian Federation: A Note" (2015) 46(3) ODIL 256, 257.

79 R Dremliuga, "A Note on the Application of Article 234 of the Law of the Sea Convention in Light of Climate Change: Views from Russia" (2017) 48(2) ODIL 128, 132.

80 The Federal Law of July 28, 2012, N 132-FZ "On Amendments to Certain Legislative Acts of the Russian Federation Concerning State Regulation of Merchant Shipping on the Water Area of the Northern Sea Route" (unofficial translation) at www.nsra.ru/en/ofitsialnaya_informatsiya/zakon_o_smp.html (accessed 5 June 2021).

81 Gavrilov, see note 78, p. 257.

82 Ibid., pp. 257–258.

83 Ibid., p. 257.

powers to regulate shipping activities occurring therein.[84] Third, to go along the route of artificially dividing the NSR into different maritime zones, whereby different legal regimes would come to apply to different parts of the NSR, is irreconcilable with the latter being an indivisible whole.[85] Its indivisibility would require that a coastal State's powers to regulate shipping must be distributed evenly across the board. This consideration trumps the tension that it creates with the navigational rights and freedoms foreign flagged vessels would usually enjoy in a foreign coastal State's territorial sea and EEZ, or on the high seas.

The 2013 Rules of Navigation in the Water Area of Northern Sea Route (2013 Rules) shine some light on the Russian Federation's position with regard shipping in the NSR.[86] Two of its main aims are to ensure safe navigation and adequate protection of the marine environment in the NSR. One of the requirements that flow from the 2013 Rules is that every vessel aiming to transit the NSR needs prior permission from the Russian authorities. Furthermore, ice breaking services may need to be hired and fees will need to be paid prior to being allowed to go through the NSR.[87] The cynically inclined observer may perhaps draw the conclusion from this that the Russian Federation's control over the NSR is mainly driven by economic interests. While there are of course economic considerations involved, that is too much of an oversimplification. For example, the 2013 Rules allow for independent navigation of certain types of icebreaking vessels and, more generally, the Russian Federation has a clear stake in keeping the Arctic environment secure and clean.[88]

The Russian Federation has not secured formal recognition that navigation in the NSR is under its complete control. In fact, the United States has opposed the Russian Federation's designation of the entire NSR as falling within the scope of its internal waters. More specifically, the United States has objected to the claiming of internal water status by the Russian Federation for certain NSR straits that are located within the Arctic Ocean (i.e., Vilkitski, Shokalsky, Laptev, and Sannikov).[89] Its opposition came to light in July 1964, after the then USSR complained about United States military vessels attempting to transit through the aforementioned straits without asking prior permission from the USSR. Almost a year later, in June 1965, the United States response came by way of a diplomatic note in which it laid out its disagreement with that position. The gist of the United States' position was that because these are straits, as they connect two parts of the high seas and are used as routes for international navigation, a right of innocent passage exists for foreign flagged vessels.[90]

11.4 The international law of the sea framework

When looking at the international legal framework applicable to the Arctic, the law of the sea will form the logical starting point.[91] Here it is recognised that, depending on the specific maritime zone involved, States have sovereignty, sovereign rights, jurisdiction,

84 Ibid., p. 258.

85 Ibid.

86 Rules of Navigation in the Water Area of the Northern Sea Route, approved by the Ministry of Transport of the Russian Federation No. 7 of January 17 (2013) (2013 Rules) at www.nsra.ru/files/fileslist/120-en-rules_perevod_cniimf-13_05_2015.pdf (accessed 5 June 2021).

87 Zou, see note 27, pp. 291, 293.

88 Dremliuga, see note 79, p. 128; Gavrilov et al., see note 25, p. 1.

89 Gavrilov, see note 78, p. 259.

90 Byers, see note 55, pp. 144–146.

91 Skodvin, see note 21, p. 153.

freedoms, rights, and obligations. Further recognition of the relevance of the law of the sea for the Arctic region emerges from the Ilulissat declaration, where the Arctic States themselves (i.e. Canada, Denmark (including the Faeroe Islands and Greenland), Finland, Norway, the Russian Federation, and the United States) identified the law of the sea as constituting the relevant legal framework.[92] Its primary regulatory instrument is the LOSC. Out of the five Arctic coastal States that have coastlines on the Arctic Ocean proper (i.e. Canada, Denmark (including the Faeroe Islands and Greenland), Norway, the Russian Federation, and the United States), only the United States is not a party. However, the United States has recognised that large parts of the LOSC reflect customary international law, whereby rights and obligations are imposed on it in that way.[93]

A first implication of the law of the sea forming the relevant legal framework is that States bordering the Arctic Ocean are entitled to claim maritime zones from their baselines: that is, a territorial sea, contiguous zone, EEZ and a continental shelf. Varying with the maritime zone involved, the coastal State has different powers, and third States are attributed rights, freedoms and obligations in these maritime zones as well. Particularly relevant in the context of Arctic shipping are the different navigational rights and freedoms that are attributed to third States when traversing through the aforementioned maritime zones. Beyond having rights, coastal States also have several obligations. Part XII of the LOSC is specifically dedicated to the protection and preservation of the marine environment, which imposes a general duty on States to regulate all sources of marine pollution, including those from shipping activities. This Part XII sets the balance between the powers and duties that coastal and flag States respectively possess in a way that those of the coastal State are enhanced, which, as a corollary, comes to the detriment of the powers of the flag State. In this vein, as discussed later,[94] with regard to ice-covered waters this balance has tilted firmly in favour of the coastal State.[95] More specifically, Part XII of the LOSC contains a provision, that is Article 234, which is of particular relevance to the Arctic Ocean. It recognises that ice-covered areas are a different kind of animal that not only requires special consideration, but that in the governance of such waters the coastal State must possess exceptional powers to prevent pollution from vessels.

11.4.1 Internal waters

Internal waters are part of a State's territory concerning which it has sovereignty. In more concrete terms, internal waters lie landward of the baselines which are used to measure the territorial sea (Article 8 of the LOSC). The precise extent of a State's internal waters interacts with the type of baseline being used, that is a normal or straight baseline. If a State is able to use the latter, it will not only create larger areas of internal waters for that State, but it will also shift its maritime zones further seawards. Both Canada and the Russian Federation have contended that the NWP and NSR are part of their internal waters. Two legal bases have been invoked by these States in support of their positions: first, that they

92 Arctic Ocean Conference, The Ilulissat Declaration, (May 28, 2008) at https://arcticportal.org/images/stories/pdf/Ilulissat-declaration.pdf (accessed 5 June 2021).

93 Kraska, see note 28, pp. 269–275.

94 Section 11.4.3.1.

95 Kraska see note 23, p. 261.

have a historic title over the waters concerned; and, second, that they are entitled to establish straight baselines along their Arctic coasts.[96] These are two issues that could easily be subjects of a paper of equal length as this one on their own. Therefore, it is only attempted here to lay out these two contentions in very broad strokes.

Irrespective of which of the two legal bases is relied upon, and assuming the exception of Article 8(2) of the LOSC does not apply to these waters, an identical effect follows: the Arctic waters in question are transformed into internal waters. This has an important implication for navigation, as there would be no internationally recognised navigational rights or freedoms for foreign flagged vessels within these waters. The difficulty with the view that, irrespective of Canada's internal waters claim being valid, there would be a right of innocent passage, because of these waters falling within the exception of Article 8(2) of the LOSC is that historically there was little international navigation in the NWP.[97] However, the contrary position has also been maintained: looking at the historic record, the amount of traffic going through the NWP has been extensive enough for a right of innocent passage to exist.[98] More generally, and to home in on Canada's historic title claim, the more commonly held view is that it would be unlikely to stand up to scrutiny if the issue were referred to international adjudication.[99] Two of the main difficulties associated with the Canadian position flow from the fact that its historic title claim is of rather recent origin, and that there would have to be acquiescence on the part of the States which are particularly affected, which is difficult to uphold considering that protests have been made by other States, including its neighbour the United States.

The alternative legal basis that has been invoked by both Canada and the Russian Federation is that they are entitled to draw straight baselines along those parts of their coasts abutting on the Arctic Ocean, which renders the waters located therein similarly internal waters. In terms of its lineage, the principles underlying the use of straight baselines can be retraced to the *Anglo Norwegian Fisheries* Case, which was decided by the International Court of Justice (ICJ) in 1951.[100] Later, in a different case, that is, *Qatar v Bahrain*, the ICJ indicated that because of their exceptional character, straight baselines have to be used "restrictively".[101] Here, the ICJ made it clear that the use of straight baselines is the exception to the general rule, which is that coastal States have to determine normal baselines along their coasts.[102] More specifically, the use of straight baselines is only allowed in two situations: that is, if the coastline is deeply indented and cut into, or if there is a fringe of islands.[103] However, there are some limitations: established straight baselines along a State's coast must follow its general direction, and the sea areas enclosed by these baselines must hold a close link to the land territory (Article 7(3) of the LOSC). On both accounts, the straight baseline system created by Canada has been questioned in academic circles.[104]

96 Pharand, see note 42, pp. 5–29.

97 Ibid., pp. 42–44.

98 Tuerk, see note 7, p. 126.

99 Pharand, see note 42, pp. 3, 13; Tuerk, see note 7, p. 124.

100 *Anglo-Norwegian Fisheries* (*UK v Norway*) (Judgment) (1951) ICJ Rep 116.

101 *Maritime Delimitation and Territorial Questions between Qatar and Bahrain* (*Qatar v Bahrain*) (Judgment) (2001) ICJ Rep 40 103 [212].

102 Ibid., [210–211].

103 Article 7(1)(2)(3)(4) of the LOSC.

104 Schofield et al., see note 8, p. 35; Krafft, see note 12, p. 537.

A similar pattern has emerged in State practice, with some States contending that either the historic title claim of Canada and/or the use of its straight baselines is not supported by the relevant international law. In 1986, the United States protested against both the drawing of straight baselines by Canada, and its historic title claim as being without foundation in international law.[105] At the root of the United States opposition was that Canada's designation of the NWP as constituting its internal waters, either because of it being allowed to use straight baselines or for historic reasons, had an identical effect: the passage rights that the United States enjoys by virtue of international law would cease to exist.[106] Later, the European Union followed with its own protest that was partially formulated along similar lines, although it more directly addressed the problematic use of straight baselines by Canada.[107]

11.4.2 Territorial sea

Within a coastal State's territorial sea, the maximum extent of which is 12 nm (Article 3 of the LOSC), the coastal State has sovereignty, which is not absolute, however. All States, whether coastal or land-locked, enjoy the right of innocent passage through the territorial sea (Article 17 of the LOSC). This right of innocent passage applies to all ships that operate on the surface; it thus does not exclusively apply to merchant vessels. This passage can be suspended, however, mainly if it is not innocent, or if its suspension is necessary for security purposes (Article 25(1) of the LOSC). A coastal State can adopt laws and regulations that are geared to ensuring the safety of navigation within a territorial sea area (Article 21 of the LOSC). However, these laws and regulations may not be concerned with "construction, manning or equipment of foreign ships", which is a limitation that *inter alia* guards against the creation of a fragmented international legal shipping regime.[108] Yet, there is an exception: that is, if a coastal State through its domestic law is giving effect to generally accepted international rules and standards. In a broad sense, this refers to the IMO conventions, of which the Polar Code is now a part as well.[109] However that may be, Canada has interpreted the reach of Article 234 of the LOSC very broadly, as extending to all waters located within the 200 nm EEZ limit.[110] This position runs into a difficulty: based on its language, the scope of application of Article 234 is confined to what can be considered the EEZ area, and any vessels operating there. Canada's argument for reading Article 234 of the LOSC at variance therewith is built on the following argument: that is, to avoid creating a situation where a coastal State's jurisdictional powers concerning vessel-source pollution are wider in the EEZ than the territorial sea, it is necessary to extend the reach of this Article 234 to the latter as well. The situation in the NSR is different, because the Russian Federation does not adopt a zonal approach to navigation within its Arctic waters. Rather, it has taken a sweeping approach to shipping, whereby the Russian Federation's control over this activity is extended to the NSR in its entirety, thus also to those parts thereof that can be considered territorial sea areas.[111]

105 Pharand, see note 42, p. 37.
106 Roach and Smith, see note 54, pp. 318–328.
107 Byers and Lodge, see note 5, p. 62.
108 Kraska, see note 28, p. 273; Tuerk, see note 7, p. 121.
109 Section 11.4.5.
110 Bartenstein, see note 10, p. 347.
111 Section 11.3.2.

11.4.3 EEZ

The concept of the EEZ is primarily designed to give coastal States a greater measure of control over the natural resources that are contained in the water column. All land territory is entitled to an EEZ, except for rocks in the sense of Article 121(3) of the LOSC. It extends at a maximum to a point located 200 nm from the baselines of the coastal State (Article 57 of the LOSC). The EEZ is a functional zone that is facultative: a proclamation is required – there is, however, no obligation for a coastal State to claim an EEZ. All States enjoy the high seas freedom of navigation in the EEZ (Article 87 of the LOSC). This freedom is made subject *to* the provisions of the LOSC (Article 58(1) of the LOSC). However, with regard to ice-covered waters, a specific legal regime is introduced in the LOSC in the shape of Article 234. The general thinking behind this provision is that the sensitivity of ice-covered waters, and their being home to sensitive marine environments, requires that the coastal State is to be endowed with far-reaching powers to control marine pollution from vessels. As a logical corollary to granting this exceptional power to the coastal State, significant limitations can be unilaterally placed on the freedom of navigation which would normally exist in EEZ areas; that is, in those EEZ areas falling outside the ambit of Article 234 of the LOSC.

11.4.3.1 "The Arctic exception": Article 234 of the LOSC

When viewed in the broader context of Part XII of the LOSC, of which Article 234 is a constituent part, the provision can be considered something of an outlier.[112] A more common thread running through this Part XII is that it seeks to limit the circumstances in which coastal States would be justified to unilaterally impose restrictions on shipping activities that go beyond the generally accepted international rules and standards. However, Article 234 of the LOSC singles out areas which are ice covered as warranting special treatment. The hazards posed by ice-covered areas to navigation, together with their sensitive marine environments, form the justification for allowing the coastal State to adopt unilateral measures that are more stringent than the generally accepted international rules and standards. This is illustrated by one of the requirements that underpins the activation of Article 234 of the LOSC: a pollution event in the area would have to lead to irreversible harm being caused to ice-covered waters. Prior to the adoption of any such environmental measures, IMO approval or consultation is not required.

It is necessary to start with an explanation of the basic elements of Article 234 of the LOSC. Looking at its language, if a coastal State's EEZ is located in an ice-covered area, this provision enables that State to take measures to prevent vessel-source pollution. To this aim, the coastal State is given both prescriptive jurisdiction and enforcement jurisdiction. More specifically, as is clear from the text of Article 234 of the LOSC, the right of the coastal State to adopt unilateral measures is geographically limited: it can "adopt and enforce non-discriminatory laws and regulations for the prevention, reduction and control of marine pollution from vessels in ice-covered areas within the limits of the *exclusive economic zone*" (italics added, YvL). While the international shipping rules and standards that flow from the IMO conventions apply similarly in areas that fall within the reach of

112 TL McDorman, "A Note on the Potential Conflicting Treaty Rights and Obligations between the IMO's Polar Code and Article 234 of the Law of the Sea Convention", in D Pharand et al. (eds), *International Law and Politics of the Arctic Ocean: Essays in Honor of Donat Pharand* (Brill Nijhoff 2015) 143.

Article 234 of the LOSC, the latter further extends the jurisdictional powers of coastal States whose coasts abut on waters that are ice covered. Then, the coastal State is allowed to take more stringent national measures than the generally accepted international rules and standards would normally allow on a unilateral basis that are directed at controlling, preventing, and reducing pollution from vessels. There are two counterbalancing elements to be found in the text of Article 234 of the LOSC, which places some limitations on the autonomy that this provision offers to coastal States concerning ice-covered waters. It provides that, first, laws and regulations enacted on the basis of Article 234 of the LOSC must be underpinned by the "best available scientific advice"; and, second, that the State adopting environmental measures must pay due regard to the navigational freedoms of third States.

Both Canada and the Russian Federation have relied on Article 234 of the LOSC for establishing specific regimes to the aim of protecting the marine environment of the Arctic Ocean from vessel-source pollution. These two regimes have in common that they go beyond the generally accepted international rules and standards. In relation to the Arctic waters lying off the coast of Canada, and while the language of Article 234 makes exclusive reference to the EEZ,[113] its interpretation has been at variance therewith.[114] The reasoning behind this is that to conclude differently would lead to the odd result that the coastal State's hands are wound tighter with regard to navigation in the territorial sea (where there is a right of innocent passage) as compared to the EEZ (where there is the freedom of navigation). The difficulty with this interpretation is that the language of the treaty provision, which can be considered as the main intention of the drafters of the LOSC, is set aside. This is then exchanged for a purposive reading of the language of the LOSC, which looks behind the text in an attempt to unearth the aim of one of its provisions, in this case Article 234. However, at the core of the rule of treaty interpretation, as set out in Article 31 of the Vienna Convention on the Law of Treaties (VCLT),[115] is to ascertain the ordinary meaning of a treaty provision. It considers that the intentions of the States concerned are best expressed through the language of a treaty provision. There are various cases of international courts and tribunals that support placing primary emphasis on the text of a treaty provision.[116] An interpretation reached by considering the ordinary meaning of the wording of the treaty provision may be supplemented by looking at its attendant circumstances, including the negotiation history of a provision (Article 32 of the VCLT). But this is a secondary means of interpretation, which would only come into play if the application of the primary means of application, establishing the ordinary meaning, leads to a result that is ambiguous or obscure. It is doubtful that to reach the conclusion that Article 234 of the LOSC is meant to exclusively apply to the EEZ, which could have been reasonably intended as is exemplified by this maritime zone being solely mentioned in the language of this provision, would qualify as a result riddled with ambiguity or obscurity.

The Russian Federation and Canada can at present invoke Article 234 as a solid legal basis for their domestic legalisation geared towards securing enhanced levels of protection against vessel-source pollution within Arctic waters. And this will likely continue to be the

113 Kraska, see note 28, p. 274.

114 Pharand, see note 42, p. 47.

115 Vienna Convention on the Law of Treaties (adopted 23 May 1969, entered into force 27 January 1980) 1115 UNTS 33 (VCLT).

116 See e.g., *Admission of a State to the United Nations* (Advisory Opinion) [1948] ICJ Rep 57, 4, 8; *Territorial Dispute* (*Libya* v *Chad*) (Judgment) [1994] ICJ Rep 6, 21–22 [41].

case in the more immediate future, as even if the ice cover continues to shrink, drifting ice will come to the fore as the main threat to shipping in the region. But what happens further down the road if the ice cover continues to shrink, making the Arctic waters accessible for longer periods of the year because of them becoming more "ice free"? Or what if the ice cover would disappear completely for most of the year, whereby these waters lose one of their main distinctive features, that is that of ice cover? Can this put the applicability of Article 234 of the LOSC in question? There seems ample reason to assume this, particularly because of the text of this latter provision. It indicates that in those areas "where particularly severe climatic conditions and the presence of ice covering such areas for most of the year create obstructions or exceptional hazards to navigation" Article 234 of the LOSC can be utilised as the basis upon which coastal States can "adopt and enforce non-discriminatory laws and regulations for the prevention, reduction and control of marine pollution from vessels".

In the academic literature, views have varied over the effects receding ice cover would have for the applicability of Article 234 of the LOSC. On one extreme of the spectrum, there is the view that even if the Arctic waters would no longer be covered by ice, its applicability would remain unaffected.[117] Underlying this argument, the gist of which is that the applicability of Article 234 is not tied to its ice cover, is the negotiation history of this provision, according to its adherents.[118] At UNCLOS III, Canada was a driving force behind the development of Article 234 of the LOSC. Its efforts must be mainly seen through the prism of Canada seeking to have the lawfulness of the AWPPA retroactively confirmed.[119] As negotiations at UNCLOS III progressed, it became apparent that the introduction of the new EEZ concept tilted too much in favour of the freedom of navigation, with which the AWPPA could not be reconciled. This created a problem for Canada and led it to invest significant energy in the creation of a special legal regime for ice-covered waters. Other States participating in UNCLOS III, despite having some sympathy for Canada's position and recognising that the Arctic waters are imbued with a certain uniqueness, wanted to guard against opening the floodgates for other States making similar broad unilateral claims to the aim of protecting the marine environment.[120] Complicating matters further was that many of the States at the UNCLOS III were satisfied with the recognition of the EEZ as such, which they felt suitably met their interests. Against this backdrop, whatever exception would materialise that would provide the AWPPA of Canada with the sought-after seal of approval, it had to be narrow in scope. Prior to tabling a proposal that made special provision for ice-covered areas, Canada sought the support of both the United States and Russia, which both were willing to give. The United States recognised that a tailor-made legal regime to deal with ice-covered waters was appropriate. With this backdrop in mind, Article 234 has been coined as "the Arctic exception".[121]

A further argument that has been advanced in the literature against the interpretation tying the applicability of Article 234 to the aspect of ice cover is that its shrinking could not have been foreseen at the time that negotiations were held at UNCLOS III.[122] Although

117 Dremliuga, see note 79, pp. 128–129; Gavrilov et al., see note 25, p. 5.
118 Dremliuga, see note 79, pp. 129–130.
119 De Mestral, see note 50, p. 115; McDorman, see note 112, p. 143.
120 Dremliuga, see note 79, p. 129; Gavrilov et al., see note 25, p. 2.
121 Kraska, see note 23, p. 260.
122 Gavrilov et al., see note 25, pp. 2–3.

it is difficult to say with certainty whether States participating therein had a clear grasp of changes occurring in the Arctic Ocean, various signs had already come to light indicating that changes in the ice cover of the Arctic Ocean were occurring. In an article from 1969, thus predating when negotiations began at UNCLOS III in 1973, Pharand already referred to a widening body of scientific opinion observing changes in the extent of the ice cover.[123] While at the time scientists were still gathering data to ascertain whether these were anomalies or structural changes, markers of change in the extent of ice cover in the Arctic Ocean did thus already exist.

To now revert to the negotiation history and the driving role that was played by Canada in the creation of Article 234 of the LOSC, which has led to the suggestion that, whatever its text provides, the intention of the drafters was to create a regime exclusively for the Arctic Ocean. An apparent difficulty with this view is that the text does not make mention of this region. Rather, both the language of the provision and the section under which Article 234 of the LOSC is collected refer to "ice-covered areas", which points in the direction that a clear connection with the physical outlook of a maritime area is established. Referring to ice-covered areas broadly, and with an ordinary reading of Article 234 of the LOSC, as Article 31 VLCT prescribes, the scope of application of this provision has to be considered as being more extensive than to solely apply to Arctic waters. Alternative candidate areas where Article 234 of the LOSC could find application are limited, but some southerly located waters could perceivably meet this definition as well.[124] However, a significant problem with the argument that Article 234 was negotiated not with the aspect of ice cover in mind, but rather to deal specifically with the Arctic Ocean, is that an ordinary reading of the language of this provision suggests something different, however.

It is here that the view lying on the other extreme of the spectrum comes to the fore: the aspect of ice cover is a key aspect for its applicability, as exemplified by the language of Article 234 of the LOSC. Due to this element being in play as a threshold requirement, and if the ice cover in an EEZ continues to shrink, the inevitable conclusion would have to be that Article 234 will lose its relevance at some point.[125] This view emphasises that the jurisdictional powers this provision grants to the coastal State are exceptional, in that they are made subject to certain conditions and limitations. One of the qualifications that due to climate change may be a threshold that is exceedingly difficult to meet (also for Arctic waters) is that the EEZ area in question must be covered with ice for most of the year, because of which challenges in the shape of navigational "obstructions or exceptional hazards" to shipping emerge. Important in this regard is that that language of Article 234 of the LOSC refers to "severe climatic conditions *and* the presence of ice" (italics added, YvL). The text thus uses not the disjunction "or", but rather the word "and", which suggests that both conditions must be present simultaneously. But can a body of water be considered to still fall within the ambit of Article 234 of the LOSC after it loses its ice cover for more than half of the year? As aforementioned, with an ordinary reading of its language, the answer would have to be: no. And yet practical questions around the applicability of Article 234 are bound to arise as to when this point is reached, including how to measure this. And, what if, for example, in one year the threshold of an area being covered long enough by ice in order

123 D Pharand, "Freedom of the Seas in the Arctic Ocean" (1969) 19(2) UTLJ 210, 223–226.

124 Gavrilov et al., see note 25, p. 2.

125 DR Rothwell, "The Law of the Sea and Arctic Governance" (2013) 107 ASILP 272, 275.

to consider it "ice covered" is not reached, but in the next year it would be?[126] Or could perhaps the prospect of drifting ice cover still be considered as "ice cover" in the sense of Article 234 of the LOSC, because it "creates obstructions or exceptional hazards to navigation", which could lead to shipping incidents that "could cause major harm or irreversible disturbance of the ecological balance"?

11.4.4 A strait or not?

Within the Arctic region there are a number of bodies of water that could potentially qualify as a strait, pursuant to Articles 37 and 38(2) of the LOSC, and in which subsequently a transit passage regime could come to exist. An example thereof is the Bering Strait, which fulfills a key role in inter-Arctic shipping, due to its strategic location, in that all inter-Arctic shipping would have to transit the Bering Strait.[127] Even though, technically, there is more than one shipping route leading through the Bering Strait, there is little doubt that it meets the geographical requirement of an international strait (it connects a part of the EEZ/high seas (i.e., Bering Sea) with another part of the EEZ/high seas (i.e., Chukchi Sea)). Because of it being regularly used as an international shipping route, the Bering Strait meets the functional requirement as well.[128] However, controversy surrounds whether the NWP and parts of the NSR meet both these requirements as well.

First to turn to the relevant international law on when a body of water qualifies as an international strait. For a strait to be governed by Part III of the LOSC, and thus for there to be passage rights for foreign vessels, it needs to have two attributes. One of these attributes is geographical in nature, the other functional. Starting with the former, a strait – being a body of water that lies between two areas of land, either continents, islands or a combination thereof – must connect one part of the high seas and EEZ and another part of the high seas or EEZ. In this vein, parts of the NSR connect a part of the EEZ/high seas with another part of the EEZ/high seas. In a geographical sense, the NWP is a bit of an anomaly, however, given that it does not neatly fit the archetype of a strait, in that there is a single body of water that lies between two parts of the high seas/EEZ. Rather, there are multiple interconnecting routes located within the NWP.[129] However, this existence of alternate routes is not an impediment to the NWP meeting the geographical requirement.[130]

The success of the argument that the Arctic shipping routes qualify, in full or in part, as international straits is more entwined with the second attribute that a strait must possess: the route has to be used for international navigation. The requirement of "international" traffic implies that the waters must be used by vessels that are flagged to different States than those adjoining the strait directly.[131] A more obvious way to determine this is by looking at the historical record of shipping and assess the degree to which that route has been used for navigational purposes.[132] But the crux is that there is little guidance on how extensive

126 Kraska, see note 36, p. 239; K Bartenstein, "The 'Arctic Exception' in the Law of the Sea Convention: A Contribution to Safer Navigation in the Northwest Passage?" 42(1–2) ODIL 22, 31.
127 Rothwell, see note 41, p. 173; Molenaar, see note 2, p. 293.
128 Rothwell, see note 4, p. 94.
129 D Pharand, *Canada's Arctic Waters in International Law* (CUP 1988) 157–158.
130 Rothwell, see note 41, pp. 171–172.
131 Pharand, see note 42, p. 30.
132 Rothwell, see note 4, p. 89.

traffic through a strait has to be. Some guidance can be drawn from the *Corfu Channel* case, however.[133] Against the backdrop of the United Kingdom contending that the geographical requirement is preeminent, and with Albania placing emphasis on the functional requirement, the ICJ stated that "[i]n the opinion of the Court the decisive criterion is rather its geographical situation as connecting two parts of the high seas and the fact of its being used for international navigation".[134] Also, the fact that the Corfu Channel was only an alternative route of passage between the Aegean and Adriatic Seas, rather than the sole existing route, and thus not being a necessary one, was not regarded as a decisive consideration by the ICJ.[135] It went on to state that the main aspect that needed to be placed at the core of the ICJ's analysis was whether it is a "useful route".[136] This would be exemplified by it being used as a route for international shipping to a certain degree.[137] High levels of international ship traffic traversing through those waters would not be necessarily required considering the ICJ's usage of the word "useful". Other questions do, however, arise, around whether all recorded transits carry equal weight in the determination of whether a body of waters satisfies the functionality requirement. In this vein, are "older" transits less relevant than more recent ones, or would uses by governmental vessels, such as submarines, carry less weight in the equation compared to commercial vessels? Or, in the broader scheme of things, is it not so much actual uses that are relevant, but rather that the body of water is susceptible to future navigational future uses?

Turning to the Arctic waters, where ice conditions have historically inhibited them from being extensively used for shipping purposes, although arguably a long history of shipping has accumulated over time in the region.[138] Given that, on the whole, international maritime traffic going through the Arctic shipping routes is on the low side, this could lead to difficulties in respect of these routes meeting the functional requirement, in order to be characterised as an international strait. In a similar vein, complicating matters as to whether the NWP satisfies the functionality requirement is that, historically, inter-Arctic passages through it have only more seldom occurred.[139] Most often coastal shipping has occurred in the NWP. Further, foreign uses of the NWP that have occurred have also been regularly in the form of submerged United States submarines travelling beneath the ice. To be more specific, since the 1960s, submarines flagged to the United States have sailed through the NWP submerged, although exact numbers are not readily available.[140] The *USS Seadragon* was seemingly the first United States submarine to sail through the NWP in a westerly direction in 1960.[141] This was followed two years later by the transit of the *USS Skate*, which crossed the NWP in the opposite direction.[142] These crossings by submarines gives rise to the question whether military uses of the NWP enhance the argument of it being used as a route for international navigation. There is no reason why submarine traffic has to be discarded from the equation. International law provides no indication that only the use by

133 *Corfu Channel Case* (*United Kingdom v Albania*) [1949] ICJ Rep 4.
134 Ibid., p. 28.
135 Ibid.
136 Ibid.
137 Ibid., p. 29.
138 Rothwell, see note 4, p. 67.
139 Chircop, see note 24, p. 201.
140 Krafft, see note 12, p. 566.
141 Pharand, see note 42, p. 37.
142 Ibid.

merchant vessels, rather than government vessels, could count towards meeting the functionality requirement. The minimum requirements are rather that the traffic has to be of the international variety, in that it concerns foreign flagged vessels, and that their destination may not be of the intra-Arctic variety. Similarly irrelevant would be whether the crossings took place submerged or above water.[143]

Once the geographical and functional requirement are both met, there is a right of transit passage for ships passing through a strait (Article 44 of the LOSC). This right has its origins in the LOSC. The right of transit passage incorporates the freedom of navigation, and when placed on a spectrum, falls somewhere between the rights of innocent passage and the freedom of navigation. Building on the ICJ's recognition in the *Corfu Channel* case that a strait is comparable to an "international highway",[144] a vessel has to proceed without delay. A vessel may transit through a strait in accordance with its "normal mode". International shipping regulations must be complied with while a vessel is engaged in transit passage (Article 39(2)(a) of the LOSC). A thorny issue is whether Article 234 of the LOSC would apply to straits as well, which is another topic that has given rise to opposite views.[145] A relevant consideration in this context is that Article 42(2) of the LOSC places emphasis on that the "practical effect" of adopted laws and regulations, which would include those based on Article 234 of the LOSC, may not amount to "denying, hampering or impairing" the right of transit passage of vessels.

11.4.5 The Polar Code

Much of the architecture of international shipping rules and regulations was not specifically designed with ice-covered waters in mind. This, at the same, lays bare their limitations when applied to areas which are imbued with more unique characteristics, which render them more susceptible to damage from shipping activities than is the case elsewhere. Against the backdrop that polar waters present an array of challenges to shipping, together with the recognition that these waters are home to pristine ecosystems, a need was recognised by the IMO for a more tailor-made response. Its efforts ultimately culminated into the Polar Code, which entered into force on 1 January 2017.[146]

The issue of designing rules and standards tailored to meet the idiosyncrasies of ships operating in polar waters emerged first on the agenda of the IMO in 2009. Discussions within the IMO bogged down on several questions, however. This was because, at the time, issues such as the scope of application of the (future) Polar Code, whether differential treatment should be given to Arctic waters and Antarctic waters respectively, and whether beyond commercial vessels provision should also be made for other types of vessels, all gave rise to divergent views. In late 2012, it was decided by the Maritime Safety Committee (MSC) that the way ahead for the creation of specific shipping rules and standards for polar waters was not to design a *de novo* treaty.[147] Rather, it was deemed more appropriate to pursue a less trouble-strewn path: that is, to make the necessary amendments to a number of IMO

143 Rothwell, see note 41, p. 163.
144 *Corfu Channel Case*, see note 133, p. 28.
145 Pharand, see note 42, p. 37; Bartensein, see note 126, pp. 33–34.
146 Bai, see note 9, p. 675.
147 VanderZwaag, see note 23, p. 301.

conventions that were already in existence. The key consideration underlying going along the route of amendment, more specifically of the MARPOL and the SOLAS, was the availability of the tacit procedures through which changes could be effected to their texts. That is, as long as the number of IMO member States registering protests does not exceed 50% of the world's shipping fleet tonnage.[148] In the first half of 2014, the rough contours of the Polar Code were already clear. However, some of its details still needed to be hammered out at that point, which delayed its adoption somewhat. The first part (I-A), which deals with safety, was adopted in November 2014 by the MSC at its 94th session. Six months later, in May 2015, the Marine Environment Protection Committee adopted the second part (II-A), which lays down pollution prevention measures.

Given that the Polar Code is part of the SOLAS and the MARPOL, it provides legally binding shipping rules and standards. One of the key aspects of the Polar Code is that the responsibility for ensuring compliance therewith does not lie with coastal States. Rather, this responsibility falls to flag States. This can be identified as one of the reasons why certain Arctic coastal States continue to maintain the position that the Polar Code only lays down minimum standards, in which light it remains necessary for them to continue to act unilaterally, through the setting of their own more stringent shipping rules in their domestic legislation, to make sure that the Arctic waters lying off their coasts are adequately protected.[149]

The Polar Code, which is composed of 12 chapters, has been hailed as heralding in a new chapter in polar shipping.[150] In terms of its scope of application, the Polar Code is not confined to the Arctic waters, but has a wider reach. It applies to polar waters generally, and thus also extends to the Antarctic region. Two main aims underpin the Polar Code: first, to ensure safe ship operations in polar waters; and, second, that the polar marine environments are adequately protected. To achieve both these goals, Part I-A of the Polar Code imposes several measures that are concerned with the issue of safety of navigation and its Part II-A lays down measures to prevent vessel-source pollution. The type of safety measures provided by the Polar Code, which relate to shipping operations generally, fall along two lines. Put broadly, the first set of rules seeks to make sure that the ships navigating through polar waters can physically withstand their icy conditions. This includes that a ship's structure must be sufficient and that the ship is watertight.[151] The second set of measures relates more broadly to the preparatory phase, which precedes undertaking a polar voyage. Prior to embarking thereon, one of the set requirements is to obtain the necessary documentation, including a Polar Ship Certificate. A further requirement is that adequate means to maintain communications while operating in polar waters need to be available on board.[152]

The other half of the measures the Polar Code provides are concerned with environmental protection. Viewed on the whole, these rules are more stringent than those that would apply in maritime areas not covered by ice. Part II-B is composed of five chapters, which are organised around the themes of oil pollution, sewage, garbage, packaged harmful substances, and noxious liquid substances.[153] Chapter 1 adopts the same standard that is

148 Article 16(2)(f)(ii) of the MARPOL; Article VIII(b)(vi) of the SOLAS.
149 Bartenstein, see note 10, p. 340.
150 Bai, see note 9, p. 675.
151 See e.g., Polar Code, see note 14, I-A Chapter 5 Watertight and Weathertight Integrity.
152 Ibid., I-A Chapter 10 Communication.
153 Bai, see note 9, p. 682.

applicable to Antarctic waters, pursuant to which vessels are prohibited from releasing any oil or oily mixtures into the polar environment.[154] In Chapter 2, which concerns the release of noxious liquid substances into the polar environment, a similar approach is followed, by imposing a blanket ban thereon.[155] Chapter 3 is a bit different compared to the other four chapters dealing with pollution prevention. By way of contrast, it does not introduce more stringent environmental protection standards for packaged harmful substances than already provided for by the MARPOL. At the root of this lies the recognition that the detrimental effects thereof are not more severe in polar waters compared to other parts of the seas and oceans, meaning that Annex III as it stands can adequately address the environmental concerns caused by releasing packaged harmful substances into polar environments.[156] Discharges of sewage is the central theme in Chapter 4. Such discharges are only allowed if they are permitted under Annex IV of the MARPOL, but otherwise they are prohibited.[157] A similar approach is pursued in Chapter 5, where the release of garbage into polar waters is permitted within limited circumstances; that is, if they take place in line with Regulation 4 of Annex V of the MARPOL.[158]

The Polar Code does not comprehensively deal with issues arising from Arctic shipping, as several gaps in its coverage remain, including that it does not address environmental threats posed by the discharge of ballast water in its mandatory provisions, nor that it prohibits heavy fuel oil use by vessels, which leads to the emission of black carbon, navigating through Arctic waters in a way similar to Antarctic waters. In this light, the adequacy of the Polar Code will remain a subject for debate for the foreseeable future. Discussion has also arisen about how the Polar Code relates to legislation adopted by Arctic States on the basis of Article 234 of the LOSC. This has given rise to divergent views, which range from that the domestic legislation adopted on that basis remains unaffected[159] to that the Polar Code has superseded marine environment legislation enacted by some of the Arctic coastal States.[160]

11.5 Conclusion

Global warming and climate change have led to the melting of the Arctic ice cover, whereby several routes have been opened, or will open in the future, that could be utilised for commercial shipping purposes. These are the NWP, the NSR and a (future) CAOR. Despite the harsh weather conditions, and the additional costs that must be borne by ship operators to ensure that their ships are able to withstand the icy conditions while transiting Arctic waters, Arctic shipping has a strong gravitational pull. The main rationale for using the Arctic shipping routes for international commercial shipping lies in its reduced transit times. More specifically, they may significantly cut the journeys of ships in terms of distance travelled, and hence reduce fuel consumption, which saves costs. One potential major advantage of transiting through Arctic waters is that two major chokepoints

154 Polar Code, see note 14, II-A Chapter 1 Prevention of Pollution by Oil.
155 Ibid., II-A Chapter 2 Control of Pollution by Noxious Liquid Substances in Bulk.
156 Bai, see note 9, p. 683.
157 Polar Code, see note 14, II-A Chapter 4 Prevention of Pollution by Sewage from Ships.
158 Ibid., II-A Chapter 5 Prevention of Pollution by Garbage from Ships.
159 Dremliuga, see note 79, p. 131; Gavrilov et al., see note 25, pp. 4–5.
160 Skodvin, see note 21, p. 157.

in international seaborne trade can be avoided: that is, the Suez Canal and the Panama Canal.[161]

More general questions remain, however, around whether the Arctic shipping routes becoming increasingly accessible will translate to them becoming a mainstay for international shipping, or that their use continues to be more incidental, particularly in the short term. Annual variability of ice cover continues to make it difficult to predict when the Arctic shipping routes are open. This unpredictability underpinning Arctic shipping is one reason why the economic pendulum has not yet swung in its favour.[162] Perhaps it might ultimately be that a picture emerges where the Arctic shipping routes will be used by vessels having a particular business model and only in relation to a more narrowly defined range of routes. To use a cliché, only time will really tell.

Nonetheless, the fact remains that it is exceedingly possible for those actors holding an interest in Arctic shipping to act upon thereon if they wish to do so, due to shipping routes being increasingly accessible because of climate change. That there is an appetite on the part of shipping companies is illustrated by the fact that the Arctic waters have been used as shipping routes by their vessels. A further reflection of this is that, on the whole, inter-Arctic ship traffic is increasing, although overall levels remain relatively low. As things currently stand, most Arctic voyages involve coastal shipping and to a much lesser extent involve ships having ports located outside of the Arctic Ocean as their destination.

An increase in shipping traffic in the Arctic Ocean, which is a prospect that looms if the ice cover continues to retreat, at the same time raises the profile of the various international legal issues that it gives rise to. And as the ice cover recedes, new life may be blown into these issues, which will also increase the need for them to be addressed in some way. There are two prominent legal issues that can be distinguished concerning Arctic shipping. The first legal issue centres on the fact that there are different views with regard to the status of the NWP and NSR. This issue is in no way new, as its origins can be retraced to the late twentieth century.[163] However, it has persisted up to the present day with little prospect of such differences being resolved in the short term. While there is little disagreement over whether the NWP and NSR connect two parts of the EEZ/high seas, and thus meet the geographical requirement, less clear is them meeting the functional requirement, in that these waters are used for international navigation. In this vein, the more limited number of international passages are the main obstacle standing in the way of the argument of the NWP and parts of the NSR qualifying as international straits under Articles 37 and 38(2) of the LOSC. One reason that underpins this lack of functional use for purposes of international navigation is more obvious and as follows: it is only when ice cover has started to recede due to climatic changes, not counting submerged vessels and coastal shipping, that uses that would qualify further towards satisfying the functionality requirement start to accelerate.

Considering the lower likelihood of the aforementioned issue being definitively resolved in the near future, it has been argued that attention should rather be directed to that ship navigation can occur safely within the Arctic Ocean.[164] It is here, however, that the second legal issue associated with Arctic shipping comes to the fore: that is, to what degree are

161 Zou, see note 27, pp. 287–288.
162 Schofield et al., see note 8, pp. 40–42.
163 Kraska, see note 28, p. 268.
164 Tuerk, see note 7, pp. 128–129.

Arctic States permitted to set requirements for shipping activities under their domestic laws that exceed the international shipping rules and standards, but which are underpinned by a desire to protect the Arctic marine environment sufficiently?

While successful international transits of the NWP and NSR are hailed with enthusiasm, together with that almost each such a new successful international voyage is interpreted as being new writing on the wall that this is going to be the "new normal",[165] there is another side to such an increase in shipping activity as well. Shipping activities and the Arctic do not necessarily go well with each other.[166] The opening of Arctic shipping routes, and a further increase in the amount of shipping that is expected to pass through them, raises concerns for the Arctic marine environment and its ecosystems. The crux lies with that these remain relatively pristine, which makes them more susceptible to the detrimental effects of shipping activities. Also, considerable variances in the density and hardness of ice, accompanied by the harsh climate of the Arctic, have the effect that navigation is a more unpleasant and a more generally risky proposition. Drifting ice, which will be an inevitable side effect of the Arctic Ocean's warming, is one particular risk that will remain, and which will only be exacerbated as the effects of climate change are increasingly felt in this region, raising the likelihood of shipping incidents with dire consequences for the marine environment occurring.[167]

Arctic waters present an array of additional hazards to shipping,[168] including strong winds, drifting ice, snowy conditions, and operating in a fragile marine environment whereof full map coverage is not available. In recognition of this, Arctic coastal States are likely to continue to point to the key role that they play in regard of the regulation of Arctic shipping. Connected to this is that they may well emphasise that taking unilateral measures will remain necessary to secure safe navigation and to limit the negative outfall of shipping activities on the marine environment.[169] This would be further necessitated by the fact that the current international legal architecture remains to fall short of what is necessary to deal with the idiosyncrasies of the Arctic environment.

The legal basis that has been invoked by both Canada and the Russian Federation to proclaim laws and regulations that are geared towards the protection of the marine environment from vessel-source pollution from commercial vessels is Article 234 of the LOSC. Nonetheless, the application of this provision may start to run into difficulties as the effects of climate change become increasingly more conspicuous. It might be still a few decades away before this argument may take actual root, but there will inevitably come a point at which Article 234 of the LOSC can no longer reasonably function as the foundation on which Arctic States can built their environmental legislation to justify greater interferences with the navigational rights and freedoms of third States, in what then could be considered no longer ice-covered areas, but "normal" EEZ areas. In the same vein, Rothwell has observed that there will come a point where Canada and the Russian Federation should re-evaluate their existing legislation based on Article 234 of the LOSC, and whether it still

165 'Historic Sea Route Opens Through Canadian Arctic Waters', see note 44.
166 Schofield et al., see note 8, pp. 40–41.
167 Zou, see note 27, pp. 287–288; Schofield et al., see note 8, p. 35.
168 RD Brubaker, *The Russian Arctic Straits* (Brill Nijhoff 2004) 2–4.
169 Byers, see note 55, 129.

matches the situation on the ground: that is, whether the Arctic waters still can be considered ice covered.[170]

A key consideration that puts the applicability of Article 234 of the LOSC into question is shrinking ice cover. Then, the question arises: how do Arctic waters differ from other waters that are similarly home to sensitive maritime environments? For example, those identified in Article 211(6) of the LOSC; this article allows the coastal State, after having obtained the approval from the competent international organisation, to adopt laws and regulations to the aim of environmental protection that go beyond the generally accepted international rules and standard within clearly defined areas of its EEZ. The answer may lie in that, besides the issue of ice cover, there is a unique sensitiveness that the Arctic environment is imbued with.[171] However, once the Arctic waters lose their ice cover, they are deprived of one of their key attributes that justified the creation of a provision dealing with ice-covered areas specifically. That this provision has a narrow scope of application is exemplified not only in that the wording of ice cover, or variant thereof, is found in the text and title of Article 234 of the LOSC, but similarly in the title of the overarching section under which this Article 234 is the only provision.

The Polar Code may be considered as a step in the right direction, also from the perspective of Arctic States, as is exemplified by the active role they played in its creation. While it may plug some of the holes that are perceived to exist by Arctic States on the international plane concerning the governance of shipping activities in the Arctic Ocean, the domestic legislation of Canada and the Russian Federation and the international legal framework continue to misalign on points. But whether these two States will be willing to bring the content of their domestic legislation in full conformity with the lower standards set on the international plane is doubtful. The key to at least part of the answer lies in the fact that Canada and the Russian Federation have shown a historical willingness to act concerning the Arctic waters if it suits their interests. Along these lines, Canada has not hesitated to act unilaterally in the name of environmental protection, and the Russian Federation has unilaterally extended its control over navigation to the NSR broadly, with both being convinced that their positions are in line with international law.[172]

A problem is that the current international law foundation on which much of the legislation of Canada and the Russian Federation is built will continue to melt away as the ice cover shrinks. There are alternative ways under international law that could perceivably be utilised by these States to introduce an adequate level of protection for the Arctic marine environment, however. But going along that route might be a path strewn with more difficulties, as it revolves around multilateralism. As various States are involved, who may have different interests and agendas, this does makes it difficult to predict whether, through such multilateral efforts, a level of protection of the Arctic marine environment can be secured that Canada and the Russian Federation deem sufficient. Despite its shortcomings, the Polar Code provides for a measure of optimism that existing gaps can be addressed on the international level, also given that the IMO has indicated that it is yet to close the

170 Rothwell, see note 125, p. 275.

171 J Ho, "The Implications of Arctic Sea Ice Decline on Shipping" (2010) 34 MP 713–715.

172 Dremliuga, see note 79, p. 132; A Chircop, "Jurisdiction over Ice-Covered Areas and the Polar Code: An Emerging Symbiotic Relationship?" (2016) 22(4) JIML 275, 285.

book on Arctic shipping.[173] All things considered, it will be some time before the ice cover has shrunk completely, or that ice in its floating form will no longer pose a threat to Arctic shipping. In the meantime, Canada and the Russian Federation can rely on the foundation of Article 234 of the LOSC to justify domestic legislation that extends beyond the generally accepted international shipping rules and standards.[174] However, as this basis will become increasingly less tenable in line with the pace at which the ice cover retracts, the preferred approach from an international legal point of view would be for these Arctic States to commit themselves more firmly to the path of multilateralism to see where it leads in terms of ensuring an adequate level of protection of the Arctic marine environment.

173 International Maritime Organization. Report of the Maritime Safety Committee on Its Ninety-Eighth Session. Doc. MSC 98/23 (June 28, 2017) 49.

174 B Oxman, "Canada's Arctic Waters: Circumnavigating the Legal Dispute" in D Pharand et al. (eds), *International Law and Politics of the Arctic Ocean: Essays in Honor of Donat Pharand* (Brill Nijhoff 2015) 198.

CHAPTER 12

Paving the way for a European Emissions Trading System for shipping

EU and IMO on different paths

Ellen J. Eftestøl and Emilie Yliheljo***

12.1 Introduction

Because of its large dependence on fossil fuels, global shipping is estimated to be responsible for around 2–3% of total global greenhouse gas (GHG) emissions, which is more emissions than any European Union (EU) State. According to the EU Commission; if the shipping sector was a company, it would rank sixth in emissions in the world.[1] The situation is even more dramatic at EU level where shipping accounted for 13% of emissions from transport.[2] According to a new study from the International Maritime Organization (IMO); the Fourth IMO GHG Study 2020, emissions from shipping will continue to increase.[3] Depending on the development in world markets related to the Covid-19 pandemic, emissions are projected to increase from about 90% of 2008 emissions in 2018 to 90–130% of 2008 emissions by 2050.

Both the IMO and the EU are constantly working on different policy instruments and corresponding legislative proposals to combat the development and reach the policy goals on a strong reduction of emissions from the transport sector. To this point GHG emissions from maritime transport has been "free of charge": The "polluter pays principle" has not applied to international shipping.[4] This position is likely to change. With the 2019 "European Green Deal" communication,[5] the Commission clearly speaks out: The price of transport must reflect the impact it has on the environment and on health.[6] The Commission will hence examine both the current tax exemptions for maritime fuels with the intent to close any loopholes, and furthermore "propose to extend European emissions trading to

* Professor, Universities of Helsinki and Oslo
** LLM, University of Helsinki

1 EU Commission 2019, Proposal for a REGULATION OF THE EUROPEAN PARLIAMENT AND OF THE COUNCIL amending Regulation (EU) 2015/757 in order to take appropriate account of the global data collection system for ship fuel oil consumption data COM(2019) 38 final at 1.

2 Ibid.

3 Smazzare, Reduction of GHG Emissions from Ships: Fourth IMO GHG Study 2020 (2020) at www.imo.org/en/MediaCentre/HotTopics/Pages/Reducing-greenhouse-gas-emissions-from-ships.aspx (accessed 20 January 2021).

4 EU Commission 2019 (n 1), ibid.

5 EU Commission 2019, 2, Communication from the Commission to the European Parliament, The European Council, The Council, The European Economic and Social Committee and the Committee of the Regions, *The European Green Deal* COM(2019) 640 Final.

6 Ibid., at 2.15.

 DOI: 10.4324/9781003155195-12

the maritime sector".[7] The EU initiative is based on a frustration with the IMO, which is considered too slow and inefficient in its response to the current climate crisis.[8] Indeed the IMO is examining different market mechanisms, which will reduce CO_2 emissions from international shipping. A CAP and TRADE system, like the EU Emissions Trading System is however, not on IMO's agenda. Instead, the organisation scrutinises whether or not a *carbon levy* for international shipping could be an effective alternative.[9]

This chapter examines and outlines the regulatory instruments already used to reduce CO_2 emissions from international shipping. The chapter takes particular interest in how *information* on emissions can be utilised as a *nudge* or for the purposes of "market-based mechanisms" to achieve the set policy goals of emission reductions.[10] Both the EU Monitoring, Verifying and Reporting (MVR) system and also the parallel IMO DCS (outlined in 4) build, among others, on the idea that collecting and reporting emissions will raise the level of environmental awareness in the industry and hence contribute to emission reductions. The systems can, hence, be seen as ways of utilising *information* as a steering mechanism. The idea behind this can be found in the *theory of nudging* (4). However, as will be outlined, the information collected by the two organisations differs. While the EU is gathering information on emitted CO_2 per transport work, voyage and ship, the IMO is collecting data on fuels spent per ship and voyage. Both organisations intend to utilise the gathered information through a "market-based mechanism" as a tool to reduce the overall emissions from the sector. Whereas the IMO is gathering data on consumed fuel from the sector and stakeholders within the IMO are examining whether a carbon levy will be an effective market-based mechanism for combatting emissions from international shipping, the EU seeks to utilise the existing emission information as foundation for an EU Emission Trading System (EU ETS) for international shipping.

12.2 European climate policy and international shipping

12.2.1 A gradual approach – striving for a global solution

European climate policy has the ambitious goal of achieving "net-zero greenhouse gas emissions by 2050 through a socially-fair transition in a cost-efficient manner". The long-term emissions vision for a climate-neutral EU was set out by the Commission in 2018 in the 2050 Long Term Strategy "A Clean Planet for All", which looks at all key sectors and explores pathways for the transition.[11] In addition to long-term targets, EU Climate Policy also comprises intermediary targets for 2020 and 2030. The targets of the 2030 Climate

7 Ibid.

8 Below in 4.1 and 5.1.

9 Below in 4.2 and 5.2.

10 As such it builds on research done by the University of Helsinki based INTERTRAN research group and particularly the authors' joint article published in Su Chen and Ulla Liukkunen (eds), *Legal Reform and the Development of Rule of Law: A Comparison between China and Finland. Promoting Sustainable Choices in Business: The Role of Emission Information* (Zhongguo fa zhi lun tan, Di 1 ban She hui ke xue wen xian chu ban she, Beijing 2019) and individual papers in E. Eftestøl-Wilhelmsson, S. Sankari and A. H Bask, *Sustainable and Efficient Transport: Incentives for Promoting a Green Transport Market* (Edward Elgar Publishing, Cheltenham UK, 2019).

11 European Commission 2018, "Communication A Clean Planet for all: A European strategic long-term vision for a prosperous, modern, competitive and climate neutral economy" [2018] COM(2018) final 773.

and Energy Framework have been worked into binding legislation, building on the legal structure of the 2020 Climate and Energy Package.[12] The overarching distinctive targets to be met on the Union level include (i) an emission reduction target, (ii) a renewable energy target and (iii) an energy efficiency target.[13] The emission reduction target is a combination of several targets.[14]

The first target is an EU-wide emission reduction target for the sectors belonging to the scope of the European emissions trading system (EU ETS), that is, energy production, energy intensive industry and civil aviation within the EU. The second target is a combined emission reduction target for all non-ETS sectors (e.g. industrial processes, agriculture and the waste sector), set in the Effort Sharing Regulation (ESR). The ESR sets binding emissions reduction targets for non-ETS sectors for each Member State.[15] The transport sector, with the exception of aviation and international shipping, is under the scope of the emissions reduction targets under the ESR. Emissions from other modes of transport than aviation and international shipping are hence calculated towards the member states' emission reduction target. The 2030 economy-wide emission reduction target, for these sectors, was increased to 55% in December of 2020.[16] International maritime transport and international aviation were, as mentioned, not included in the 2030 Policy Framework emission reduction target; however the level of ambition has grown and the EU is seeking ways to include also shipping into its emission reduction targets.[17] The process of including international shipping in EU climate policies started already with the 2011 Roadmap to a Single European Transport Area. The Roadmap set the following target for international shipping: "The environmental record of shipping can and must be improved by both technology and better fuels and operations: overall, the EU CO_2 emissions from maritime transport should be cut by 40% (if feasible 50%) by 2050 compared to 2005 levels."[18]

Despite setting regional targets, the EU has consistently prioritised and emphasised an international solution to GHG from international shipping urging the IMO to adopt measures.[19] However, because of what has been regarded as a slow response from the IMO, the EU has since 2013 prepared for ways of including CO_2 emissions from shipping in its overall emission reduction targets by publishing a strategy for integrating maritime transport

12 K. Kulovesi and S. Oberthür. "Assessing the EU's 2030 Climate and Energy Policy Framework: Incremental Change toward Radical Transformation?" [2020] Review of European, Comparative & International Environmental Law 29, no. 2: 151, at 152.

13 The emission reduction target is implemented through Directive (EU) 2018/410 amending Directive 2003/87/EC on the EU emissions trading system (ETS Amending Directive) [2018] OJ L76/3 and the Regulation (EU) 2018/842 on binding annual greenhouse gas emission reductions by Member States from 2021 to 2030 contributing to climate action to meet commitments under the Paris Agreement and amending Regulation (EU) No. 525/2013 (ESR) [2018] OJ L156/26. The renewable energy target under Directive (EU) 2018/2001 on the promotion of the use of energy from renewable sources (RED II) [2018] OJ L328/82; and the energy efficiency target under Directive (EU) 2018/2002 of the European Parliament and of the Council of 11 December 2018 amending Directive 2012/27/EU on energy efficiency (Energy Efficiency Directive) [2018] OJ L328/210.

14 European Council, Conclusions (23 and 24 October 2014) [2014] EUCO 169/14, para 1.

15 The ESR covers in other words energy from fuel combustion and fugitive emissions from fuels, which are outside the scope of the EU ETS.

16 European Council, Conclusions of meeting (10 and 11 December 2020) [2020] EUCO 22/20, para 12.

17 European Council 2014 (n 14), para 1.

18 EU Commission, White Paper Roadmap to a Single European Transport Area – Towards a competitive and resource efficient transport system [2011] COM (2011) 144 final, para 29.

19 Ibid. and more recently EU Commission 2018, (n 11) section 3.

emissions in the EU's GHG emission reduction policies.[20] In the strategy, the EU Commission reinforced the commitment to global action and ensuring "across the board" emission reductions while maintaining a global level playing field for the shipping industry,[21] while setting out a gradual approach to include maritime transport into the European climate targets.

As a first step and in response to the continuing absence of a global framework, union-wide rules for MRV GHG emissions from shipping were adopted in 2015 through the so-called *MRV Shipping Regulation*.[22] The political debate related to the regulation was inflamed and the proposal went through 80 amendments in the EU Parliament before being accepted.[23] The Regulation entered into force in January 2018, and GHG emissions from intra-EU voyages, incoming voyages from a non-Union port to a port within the Union, as well as outgoing voyages from a Union port to a non-Union port are to be monitored, verified and reported, irrespective of which flag the ships sail under.[24] The MRV framework was intended to function as a model for a global mechanism,[25] and it was successful in that the regulation proposal speeded up international efforts. In 2016, IMO adopted mandatory data-collection provisions for the fuel consumption of ships.[26]

12.2.2 Raising ambition and growing urgency – the European Green Deal and the 2030 Climate Action Plan

Despite the efforts undertaken, the results in the shipping sector have been unsatisfactory as GHG emissions from transport are not decreasing and the trend is actually the reverse. The growing urgency to curb emissions from maritime transport is recognised in the European Green Deal from 2019,[27] which also puts forward proposals for measures to reduce emissions from the sector. The Green Deal is an ambitious policy document. The final goal is that of the 2050 Long Term Strategy, that is, to transform the EU into a society with no net emissions of greenhouse gases in 2050 and where economic growth is decoupled from resource use.[28] The Green Deal is, furthermore, an integral part of the Commission's strategy to implement the United Nations' 2030 Agenda and the sustainable development goals.[29] For transport this means a *90% reduction of greenhouse gas emissions by 2050* and all modes of transport need to contribute to the reduction, including international

20 EU Commission, Communication from the Commission to the European Parliament, the Council, the European Economic and Social Committee and the Committee of the Regions: Integrating maritime transport emissions in the EU's greenhouse gas reduction policies, [2013] COM (2013) 479 final.

21 Ibid., at 4–5.

22 Consolidated Version of Regulation (EU) 2015/757 of the European Parliament and of the Council on the monitoring, reporting and verification of carbon dioxide emissions from maritime transport, and amending Directive 2009/16/EC (MRV Shipping Regulation) [2015] OJL 123/55.

23 www.europarl.europa.eu/legislative-train/theme-resilient-energy-union-with-a-climate-change-policy/file-monitoring-maritime-transport-ghg-emissions (accessed 5 September 2020).

24 Subject to a threshold for small emitters and exemption of certain vessels fish-catching ships.

25 MRV Shipping Regulation (n 22) recital 34.

26 IMO Marine Environment Protection Committee Resolution MEPC.278(70): Amendments to the Annex of the Protocol of 1997 to Amend the International Convention for the Prevention of Pollution from Ships, 1973, as Modified by the Protocol of 1978 Relating Thereto Amendments to MARPOL Annex VI (Data collection system for fuel oil consumption of ships) [October 2016].

27 Green Deal (n 5).

28 Ibid., 2.

29 Ibid., 3.

shipping.[30] With the Green Deal the Commission is proposing to implement the third step of the strategy for progressively integrating maritime emissions into EU climate policy; developing further measures, including market-based measures, in the medium to long term.[31] The choice of instrument is the inclusion of international shipping in the EU ETS.

Including shipping in the EU ETS is clearly expressed in the 2019 European Green Deal. This was, however, not part of the proposed 2019 revision of the MRV Shipping Regulation 2015/757 until the proposal was discussed in the European Parliament. Here the ENVI Committee appointed Jutta Paulus (Greens/EEA, Germany) as rapporteur for the file. Her draft report of 24 January 2020 took a radical line, and proposed to include maritime shipping in the EU ETS. Among other things, she claimed that although it is important that the EU and IMO reporting obligations are aligned, it does not mean that EU standards should be lowered. On the contrary, better standards at global level are necessary. To give an example: In contrast to the EU MRV, the IMO is not collecting data on the cargo carried but on deadweight tonnage only. This measurement makes it very difficult, if not impossible to calculate the carbon footprint of the shipped goods. According to Paulus the IMO has promised for more than 20 years that it will tackle shipping emissions and has only introduced its DCS after the EU has implemented the MRV Regulation. No real progress has been seen, and Paulus finds it necessary that the EU takes action to achieve the Paris objective to limit the temperature increase to 1.5°C above pre-industrial levels. The report furthermore expresses that although collecting data on emission is important, now is the momentum to actually use the collected data.

In September of 2020 the Commission published the 2030 Climate Action Plan on how to raise ambitions for 2030 in order to meet the climate neutrality target for 2050.[32] It contains a proposal to change the 2030 target to 55%, previews a set of actions required across all sectors of the economy and launches the revisions of the key legislative instruments to achieve the increased target. As already noted, the 2030 Action Plan states that emission reductions are needed from all transport sectors, including the waterborne transport sector. In order to achieve the increased EU-wide emission reduction target, the 2030 Action Plan calls for the maritime sector to scale up efforts to improve efficiency of ships and operations, increase the use of sustainably produced renewable and low-carbon fuels and for technology development and deployment to occur already by 2030. The EU Commission is assessing these in the Fuel EU Maritime initiative, which aims to increase the production and uptake of sustainable alternative fuels for shipping.[33] The proposed update of the 2030 Climate and Energy Framework includes reinforcing and increasing the role for emissions trading and energy taxation, that is, economic incentives for emission reductions.[34] The Commission proposes to extend the EU ETS to all combustion of fossil fuels, also in the transport sector.

---

30 Ibid., section 2.1.5.

31 EU Commission, *Final Report from the Commission – 2019 Annual Commission Staff Working Document. Full-length report: Accompanying the document Report on CO2 Emissions form Maritime Transport* COM(2020) 3184 final.

32 EU Commission, Communication from the Commission to the European Parliament, the Council, the European Economic and Social Committee and The Committee of the Regions- Stepping up Europe's 2030 climate ambition Investing in a climate-neutral future for the benefit of our people (2030 Climate Action Plan) [2020] COM/2020/562 final, 2.

33 Ibid., 9–10.

34 Ibid., 13.

Shipping is mentioned separately and recognising the trend with growing emission the Commission outlines that at least intra-EU shipping should be included in the EU ETS. The desirability of international co-operation under IMO in relation to shipping is repeated, but simultaneously the Commission states that

> it will give fresh political consideration to the international aspects of the EU ETS, taxation and fuel policies for aviation and maritime to ensure the gradual decarbonisation of all fuel use from transport relating to the EU with the ambition to include international emissions from aviation and navigation into the EU ETS".[35]

A proposal for the revision of the EU ETS Directive is expected by June 2021.[36]

12.3 Emission information as a tool to combat climate change

12.3.1 Information as a steering mechanism

As discussed in the previous section, the current instruments regulating GHG from international shipping build on information. The mitigation of climate change and reducing GHG emissions from all sectors of society requires policy action on multiple political and societal levels and raises questions of the optimal choice of policy instruments to achieve the emission reduction targets.[37] Information is, from a theoretical point of view, long recognised as a steering instrument.[38] Informational regulation is generally among the least interventionist, either prohibiting the provision of false or misleading information to the public or requiring disclosure of specific information to the public or, for example, to specific government officials.[39] Production and disclosure of information can lead to better decision-making, increase transparency and accountability as well as trigger social, political and market responses that eventually lead to better environmental outcomes.[40] Regulation applying informational steering allows market participants such as consumers or investors to make informed decisions about the acceptability of a product or services, for example in terms of their carbon footprint.[41] Informational steering has, for example.

35 Ibid., 16 The Commission also lists other instruments such as an updated methodology to promote the use of renewable and low-carbon fuels in the transport sector set out in the Renewable Energy Directive.

36 Ibid., 25.

37 See e.g. D. Benson and A. Jordan, "Climate Policy Instrument Choices" in D. A. Farber and M. Peeters (eds), *Climate Change Law* (Elgar Encyclopedia of Environmental Law, Edward Elgar, 2016), 57.

38 The question is one of choice of appropriate tools from a toolkit. The classification and typology of policy instruments varies in literature between scholars. Kokko e.g. refers to regulatory instruments of environmental regulation as instruments based on (1) informational steering (such as certificates), (2) economic steering (such as taxes or emissions trading) and (3) legal steering and other steering (such as environmental permits or prohibitions). See K. Kokko, *Ympäristöoikeuden Perusteet: Yleiset Opit, Sääntely Ja Ratkaisun Teoria* (Edita Publishing Oy, 2017), 256–262. Baldwin et al. divide the available regulatory strategies or techniques available to (governmental) regulators in categories based on (1) command-and-control, (2) incentive-based regimes (such as taxes), (3) market harnessing controls (including emissions trading), (4) disclosure regulation (supply of information) and, finally, (5) direction action and design solutions undertaken by the state. See R. Baldwin, M. Cave and M. Lodge, *Understanding Regulation: Theory, Strategy, and Practice* (2nd edn Oxford University Press), 105–136.

39 Baldwin et al. (n 38), 119.

40 B. C. Karkkainen, "Information as environmental regulation" in L. Paddock, R. Glicksman and N. Bryner (eds). *Decision Making in Environmental Law* (Elgar Encyclopedia if Environmental Law Vol. II, Edward Elgar 2016), 199.

41 Baldwin et al. (n 38).

been used in the road transport sector to complement binding CO_2 standards in the form of car labels with information intended to consumers on fuel consumption and CO_2 emission of new cars.[42]

Information asymmetries between market actors is, alone, one of the main market barriers for improvements in energy efficiency.[43] In the shipping industry a potential for cost-efficient energy efficiency improvements has been identified, but the energy efficiency improvements have been blocked by the absence of additional incentives to prevent market failures such as the lack of financing for technological improvements and, prior to the adoption of the MRV Shipping Regulation, also reliable information on, for example, vessels fuel consumption (and indirectly GHG emissions).[44] There are studies that show that lack of reliable information on GHG emissions on a vessel level has constituted a barrier for integrating energy efficiency concerns into transactions involving vessels between owners and vessel users (ship operators, charters etc.), contributing to the absence of incentives for ship owners to invest in energy efficiency improvements.[45]

While regulatory strategies based on information and disclosure have the potential of generating cost-efficient GHG emission reductions, especially when combined with more stringent and coercive command-and-control measures, there are limitations as to what can be achieved with informational regulation. In addition, there are certain requirements and limitations in relation to the information itself that affect how well information functions as a steering mechanism. One limitation is the *potentially high costs of producing information*, which could be very high due to various factors.[46] In addition, informational policy instruments contain the risk of false or misleading information provided to the market and requires some level of policing of information. In some cases, the risk or damages from a product or activity might simply be so severe that disclosure regulation alone is not sufficient and requires more coercive regulation as a compliment.[47]

From the viewpoint of informational steering and its efficiency, a central element is furthermore *how the information is received and affects market participants*. The information disclosure might, for example, require some standard setting by regulators to promote the disclosure of information in a format that is informative and supports decision-making. There might also be challenges in relation to the characteristics of the users or recipients of the information and pure mistakes made by the previously mentioned, in their use of information or failures in understanding the implications of information given. There is also a risk that users do not respond to the information as anticipated by the regulator, for example by continuing to choose based solely on price rather than environmental impact of products. If this is the case, attending to informational asymmetries in relation to the availability of information might not be sufficient. The next part will therefore explore how

42 Directive 1999/94/EC of the European Parliament and of the Council of 13 December 1999 relating to the availability of consumer information on fuel economy and CO_2 emissions in respect of the marketing of new passenger cars [1999] OJ L 12/16.

43 A. Bowen and S. Fankhauser, "Good Practice in Low-carbon Policy" in A. Averchenkova, S. Fankhauser and M. Nachmany (eds), *Trends in Climate Change Legislation* (Edward Elgar, 2017), 131.

44 EU Commission 2013 (n 20) section 1.

45 Maddox Consulting, *Analysis of Market Barriers to Cost Effective GHG Emission Reductions in the Maritime Sector* (European Commission 2012), 65.

46 Baldwin et al. (n 38), at 120.

47 Ibid., at 120.

behavioural insights can be used to mitigate some of these risks and increase the effect of informational steering.

12.3.2 The theory of nudging

How behavioural insights can be exploited to tackle environmental problems has interested both academic scholars[48] as well as international organisations and decision-makers for some time. The EU, the Nordic Council of Ministers (Nordic Council) and the Organisation for Economic Co-operation and Development (OECD) have recently published reports on the use of behavioural science to promote environmental change. Why the international organisations and decision-makers are interested in this is reflected in the title of the OECD report, "Tackling Environmental Problems with the Help of Behavioural Insights" from 2017.[49] With the growing challenges of climate change all tools available to reach the common goals on emission reductions must be considered. The OECD addresses a variety of questions related to how behavioural insights can help overcome environmental problems. The idea of the so-called *homo economicus*, who, fully informed, make rational choices, is challenged by an understanding of the fact that people are not always rational. Individual decision-making is often based on framing and reference points, altruism and heuristic methods.[50] Often people do not make deliberate choices at all, but rely on shortcuts and habits.[51] In economic literature, this is characterised as deviations from rational decision-making, and hence labelled as *behavioural biases* or *limitations*. The insights of this, often addressed as *behavioural insights*, are based on a multitude of studies on human behaviour from different sciences, such as psychology, economics, sociology and neuroscience, among other sciences.[52]

One such behavioural insight is that people can be *nudged towards a certain behaviour, while still leaving them with a free choice*. In others words, it is possible to influence the way people behave, without forcing them to behave in certain way. It was the behavioural economist Richard Thaler and law school scholar Cass Sunstein who invented the term nudge in their book: *Nudge: Improving Decisions About Health, Wealth and Happiness* from 2008.[53] Thaler and Sunstein define nudges as any aspect of the decision environment "that alters people's behaviour in a predictable way without forbidding any options or significantly changing their economic incentives".[54] The idea is, in other words, that by implementing small changes of different kinds, people's decisions can be affected and turned in the right way. *The interesting part is that a behavioural change is made while the freedom of choice is retained with the person who takes the decision.* According to Thaler and Sunstein, there are several ways to present a choice to the decision-maker and what the

48 An overview is provided by M. Lehner, O. Mont and E. Heiskanen, "Nudging – A Promising Tool for Sustainable Consumption Behaviour?" (2016) 134 Journal of Cleaner Production 166, in their literature review on research on nudging as a tool for sustainable consumption behaviour.

49 Cornago et al., *Tackling Environmental Problems with the Help of Behavioural Insights* (OECD Publishing, Paris).

50 Ibid., p. 23.

51 Lehner, Mont and Heiskanen (n 48), at 1.

52 J. Sousa et. al. (2016), p. 9.

53 R. H. Thaler and C. R. Sunstein, *Nudge: Improving Decisions About Health, Wealth, and Happiness* (Penguin Books; Yale University Press, London, 2008).

54 Ibid., at p. 8.

decision-maker chooses depends in many cases upon how the choice is presented.[55] In this perspective, policy makers are seen as "choice-architects" who facilitate the way people make choices. Nudging as a policy tool is hence also described as "choice architecture". Another way of describing the phenomena is to use the less attractive term manipulating. However, nudging as a form of libertarian paternalism, where people are steered in directions that will promote their own welfare and common good is considered both possible and legitimate for private and public institutions to affect behaviour while also respecting freedom of choice.[56]

Nudges are particularly interesting in relation to the EU transport policy on sustainable transport as the EU policy rests on the notion of *freedom of choice*. So when, on one hand, a behavioural change is wanted, but, on the other, freedom of choice should be protected[57] – and business as usual is not considered an option[58] – the legislator has to be creative and look for new possibilities in order to reach the stated policy goals. In our case, these are radical goals on major reductions of emissions from maritime transport. Nudging is in this context a fascinating policy tool that should be recognised and integrated in the legislative process to a larger degree than what we have seen so far. Indeed nudges are studied mostly in relation to so-called business-to-consumer, or B2C, relationships.[59] However, firms can also be subject to nudges.[60] Firms are not theoretical entities or "black boxes".[61] Human beings, individual managers, boards or employees make decisions in firms. Hence, decision-making in firms can also be biased. Decisions in firms are, in other words, not strictly rational.[62] Accordingly, also decision-making in firms can be nudged with the right incentives. Despite the fact that *information*, due to the "human factor" not always guarantees the wanted behaviour, it clearly has an impact on people's behaviour. It is, for example, unquestionable that information on emissions do have an impact on the choices market actors make. Studies show that companies that measure and report their carbon footprints can reduce their emissions by 10–15%.[63] Information is, in other words, a powerful tool in the struggle for a behavioural shift in the transport industry and is hence an essential part of the nudging "package".[64] Presented in an optimal manner, *information is* understood to be a "promising tool to increase pro-environmental choices".[65] Indeed, the information will only have strong effect when presented in an accessible manner. It is hence indispensable

55 E. J Johnson and others, "Beyond Nudges, Tools of a Choice Architecture" (2012) 23 Mark Lett 487, p. 1.

56 Lehner, Mont and Heiskanen (n 48) pp. 174–175.

57 EU Commission, 2011 (n 18) p. 13.

58 Ibid., at 13, p. 4.

59 See e.g. H.-W. Micklitz, L. A. Reisch and K. Hagen, "An Introduction to the Special Issue on 'Behavioural Economics, Consumer Policy, and Consumer Law'" (2011) 34 J Consum Policy 271.

60 M. Armstrong and S. Huck, "Behavioral Economics as Applied to Firms: A Primer" (2010) No. 2937 CESifo Working Paper Series.

61 A. Heinmann, "Behavioral Antitrust A 'More Realistic Approach' to Competition Law" in K. Mathis (ed), *European Perspectives on Behavioural Law and Economics* (Economic Analysis of Law in European Legal Scholarship, Springer International Publishing, Cham 2015) p. 214.

62 Ibid., at p. 212.

63 SWD(016) 244 final. Commission Staff working document accompanying the white paper: A European Strategy for Low-Emission Mobility COM(2016)501 final at p. 125. The study refereed in the working document was performed by Dutch "Lean & Green".

64 F. Ölander and J. Thøgersen, "Informing Versus Nudging in Environmental Policy" (2014) 37 J Consum Policy 341, p. 354.

65 A. S. E. Nielsen, H. Sand and P. Sørensen, *Nudging and Pro-Environmental Behaviour* (TemaNord, Nordic Council of Ministers, Copenhagen 2017).

to consider both what *kind of information* is needed as well as how this information should be presented to the market actors to have the wanted effect.[66]

Information production, collection and disclosure requirements are present also in other steering mechanisms than informational policy instrument. Examples of this are environmental permits or, for example, *emissions trading*.[67] In the case of the EU ETS, the policy instrument can be categorised as a market-based steering mechanism relying on administrative steering,[68] but as will be further discussed below the environmental integrity of the instrument relies on accurate data on GHG emission from various sources. This information collected under the scheme for compliance purposes has had the effect of affecting decision-making by market actors.

As regards the wanted behavioural change towards sustainable transport, emissions standards have been a central policy instrument in the road sector for cars and vans.[69] The standards have included reporting obligations on vehicles for member states and in conjunction with the new standards for post-2020 the Commission is looking to strengthen the impact of the standards by market surveillance and collection of real time data.[70] A regulation for the monitoring and reporting of GHG emissions from heavy duty vehicles has also been adopted, obliging member states and manufacturers to monitor emissions from heavy duty vehicles registered in the Union. The monitoring is to be performed by manufacturers using the standardised Vehicle Energy Consumption Calculation Tool (VECTO). The information is published annually in the form of performance of the fleet of the Union, member states and manufacturers in terms of average fuel consumption and GHG emissions.[71]

The emissions described in the examples above are reported either to authorities or as general product information in the market. One idea is that the obligation to monitor and report emissions will increase the environmental awareness in the transport industry and the market at large, and eventually *nudge* the market actors to a more environmentally friendly behaviour and hence emission reduction. As mentioned, studies show that companies that measure and report their carbon footprints can reduce their emissions by 10–15%.[72] Presented in an optimal manner, *information is* understood to be a "promising tool to increase pro-environmental choices.[73]

However, sometimes nudging is not enough and stronger steering mechanisms are required. By introducing rules on monitoring, verifying and reporting CO_2 emissions from international shipping (below in 12.4), the EU is paving the way for an emissions trading system for the industry. The IMO has, on the contrary, so far been reluctant in following this path, but is introducing technical and operational measures to reduce CO_2 emissions. Information on

66 Ölander and Thøgersen (n 64) p. 354.

67 Karkkainen (n 40), p. 199.

68 Kokko (n 38) pp. 259–260.

69 EU Commission 2017: Proposal for a REGULATION OF THE EUROPEAN PARLIAMENT AND OF THE COUNCIL setting emission performance standards for new passenger cars and for new light commercial vehicles as part of the Union's integrated approach to reduce CO2 emissions from light-duty vehicles and amending Regulation (EC) No. 715/2007 (recast) COM(2017) 0676 final.

70 Ibid.

71 Regulation (EU) 2018/956 of the European Parliament and of the Council of 28 June 2018 on the monitoring and reporting of CO2 emissions from and fuel consumption of new heavy-duty vehicles, [2018] OJ L 173/1.

72 SWD(016) 244 final. Commission Staff working document accompanying the white paper: A European Strategy for Low-Emission Mobility COM(2016)501 final at p. 125. The study refereed in the working document was performed by Dutch "Lean & Green".

73 Nielsen, Sand and Sørensen (n 65).

fuel consumption is gathered to nudge the industry towards greener operations. This effect might be strengthened by an international carbon levy established and led by the IMO, a levy that would also give room for needed research and development.

12.4 Informational steering of GHG emissions from shipping – comparing approaches

12.4.1 The EU MRV Shipping regulation

As pointed out above, the European Union has paved the way for an emissions trading system for shipping by introducing the MRV Shipping Regulation, which entered into force in January 2018.[74] Subject to the regulation, GHG emissions from intra-EU voyages, incoming voyages from a non-Union port to a port within the Union, as well as outgoing voyages from a Union port to a non-Union port, are monitored, verified and reported, irrespective of which flag the ships sail under. The regulation is in other words *flag blind.* All ships become subject to obligations under the MRV Shipping Regulation upon entrance to a port in the jurisdiction of a Member State.[75] The obligation only applies when the ship stops to load or unload cargo or to embark or disembark passengers.[76] Corresponding to the cost-efficiency principle[77] the MRV Shipping Regulation applies to ships above 5,000 gross tonnage in respect of CO_2 emissions released.[78] Smaller ships are exempted from the MRV obligations.[79]

If subject to the MRV Shipping Regulation, the company operating the ship,[80] must monitor, verify and report annual CO_2 emissions and other relevant information arising from their ships' voyages during a reporting period, which is normally one year.[81] Both the monitoring and the reporting must be complete and cover CO_2 emissions from the combustion of fuels, while the ships are at sea as well as at berth. The regulation emphasises that the information must be reliable and accurate. A robust ship-specific Union MRV system should be based on the calculation of emissions from fuel consumed on voyages to and from Union ports, as fuel sales data could not provide appropriately accurate estimates for the fuel consumption within this specific scope, due to the large tank capacities of ships.[82] The obligation to monitor the system started in 2017 with the preparation of a monitoring

74 Above in 2.1 (n 22).

75 Ibid., art. 2.

76 Ibid., art. 3 (b); "[S]tops for the sole purposes of refuelling, obtaining supplies, relieving the crew, going into dry-dock or making repairs to the ship and/or its equipment, stops in port because the ship is in need of assistance or in distress, ship-to-ship transfers carried out outside ports, and stops for the sole purpose of taking shelter from adverse weather or rendered necessary by search and rescue activities are excluded."

77 https://ec.europa.eu/clima/policies/transport/shipping_en (accessed 18.02.2021).

78 MRV Shipping Regulation (n 22) art. 2.

79 The Exemption from the regulation can be extracted from art. 2, which holds the scope of the regulation. According to art. 2.2 the Regulation does not apply to warships, naval auxiliaries, fish-catching or fish-processing ships, wooden ships of a primitive build (!), ships not propelled by mechanical means or government ships used for non-commercial purposes.

80 MRV Shipping regulation (n 22) art. 4. According to the regulation art. 3 (d), the company equals "the shipowner or any other organisation or person, such as the manager or the bareboat charterer, which has assumed the responsibility for the operation of the ship from the shipowner".

81 Ibid., art. 9 and 11.

82 Ibid., preamble at 17.

plan.[83] The monitoring plan should be filled out by the ship owners and explain how they intend to monitor the relevant parameters required by the MRV Shipping Regulation. From 2018 onwards, companies are required to monitor CO_2 emissions from their vessels by applying the "appropriate method" for determining CO_2 emissions. Shipowners can choose between four methods, as explained in Annex 1, Part A, to monitor CO_2 emissions:

1) Bunker Fuel Delivery Note (BND) and periodic stocktakes of fuel tanks
2) Bunker fuel tank monitoring on board
3) Flow meters for applicable combustion processes
4) Direct CO_2 emission measurements.

For each method, companies have to indicate the corresponding level of uncertainty. According to the 2019 Annual Report on CO_2 Emissions from Maritime Transport[84] all companies relied on the first three monitoring methods during the first reporting period, while the alternative fourth, direct CO_2 emission measurements, was not used. As regards the uncertainty associated with fuel monitoring, the companies relied on default values following the guidance established by the European Sustainable Shipping Forum (ESSF).[85]

After having the monitoring plan assessed by an accredited verifier, the shipowners should *monitor and report* the different parameters and prepare an emission report. This should be done in an electronic inspection database called THETIS.[86] THETIS is developed, maintained and hosted by the European Maritime Safety Agency (EMSA). EMSA has developed a new module in THETIS, namely THETIS-MRV, enabling companies responsible for the operation of large ships using EU ports to report their CO_2 emissions under the MRV Shipping Regulation. THETIS-MRV includes a mandatory and a voluntary module. Through the mandatory module, companies will generate Emission Reports, which will be assessed by Verifiers who will issue an electronic Document of Compliance in the system. Through the voluntary module, companies may draft their monitoring plans and the system will make them available for verifiers' assessment.[87]

The intention behind the EU MRV Shipping regulation was that it should work as a pilot or a model for a global mechanism.[88] The process within the EU has certainly speeded up the process within the IMO, which in 2016 adopted the IMO DCS for fuel consumption of ships.[89] Indeed the interaction with the IMO influenced the EU, which in 2019 proposed an amendment of the EU MVR Shipping Regulation to adapt to the IMO DCS. The proposed revision aims to facilitate the coinciding implementation of the two systems, while preserving the objectives of the current EU legislation, particularly to keep the monitored and verified CO_2 emissions data at individual ship level, but also to stimulate innovations and energy efficiency

83 Ibid., art. 6.

84 EU Commission 2020 (n 31) at p. 14.

85 European Sustainable Shipping Forum, 'Guidance/Best practices document on monitoring and reporting of fuel consumption, CO2 emissions and other relevant parameters pursuant to Regulation 2015/757 on monitoring, reporting and verification emissions from maritime transport' (2017).

86 The name derives from the Greek goddess of the sea in mythology.

87 The system has been available from 7 August 2017 and can be reached at https://mrv.emsa.europa.eu.

88 MRV Shipping Regulation (n 74) recital 34.

89 IMO Marine Environment Protection Committee Resolution MEPC.278(70): Amendments to the Annex of the Protocol of 1997 to Amend the International Convention for the Prevention of Pollution from Ships, 1973, as Modified by the Protocol of 1978 Relating Thereto Amendments to MARPOL Annex VI (Data collection system for fuel oil consumption of ships), October 2016.

solutions. The proposal furthermore aims at reducing the administrative burden and associated costs for ships that have to report under both systems. The proposal is still pending.[90]

12.4.2 The IMO process

The IMO DCS for fuel oil consumption of ships entered into force on 1 March 2018. Under the framework, ships of 5,000 gross tonnage and above are required to *collect* consumption data for each type of fuel oil they use, as well as other, additional, specified data including proxies for transport work. According to the IMO DCS the collected data should be *reported* to the flag State after the end of each calendar year and the flag State, having determined that the data has been reported in accordance with the requirements, should *issue a Statement of Compliance* to the ship. Flag States are required to subsequently transfer this data to an *IMO Ship Fuel Oil Consumption Database*.[91] IMO will be required to produce an annual report to the Marine Environment Protection Committee (MEPC), which is the body within IMO responsible for environmental issues including climate change. Furthermore, each ship should have a Ship Energy Efficiency Management Plan (SEEMP), which must include a description of the methodology that will be used to collect the data and the processes that will be used to report the data to the ship's flag State.

Whereas the EU MRV system and the IMO DCS overlaps in many aspects, particularly as regards the ways of measuring the CO_2 emissions, the systems are totally divided when it comes to how the emission information is utilised. While the emission information collected by the EU is made publicly available and ready to be used in a potential EU ETS, the information gathered by IMO is *confidential and not publicly available*.[92] According to the IMO-system, governments that have ratified MARPOL Annex VI shall have access to the data of all ships in an *anonymised format* strictly for their analysis and consideration.[93] It is hence rather clear that the IMO is not preparing for an international CAP and TRADE solution as a market-based instrument to reduce emissions from shipping, but keeps its focus on technical and operational measurements. This is, however, the third and next step in the above-mentioned EU strategy for emission reduction from international shipping.

12.5 Prospects for market-based measures for shipping and the role of information

12.5.1 The proposed EU ETS

The Commission is currently preparing a proposal for inclusion of shipping into the EU ETS, that is, the third step of its strategy to include shipping in its emission reduction

90 EU Commission 2019 (n 1) at 1 and www.europarl.europa.eu/legislative-train/theme-resilient-energy-union-with-a-climate-change-policy/file-revision-of-the-eu-system-to-monitor-report-and-verify-co2-emissions-from-ships (accessed 18 January 2021).

91 IMO Ship Fuel Oil Consumption Database has been launched as a new module within the Global Integrated Shipping Information System (GISIS) platform and that Member States now have access to the Database. www.imo.org/en/OurWork/Environment/PollutionPrevention/AirPollution/Pages/Data-Collection-System.aspx (accessed 6 September 2020).

92 MEPC 68/21, para 4.12.

93 Ibid. The new regulation 22A states that the IMO Secretariat would provide an annual report on the data collected to MEPC.

policies. Emissions trading is a market-based mechanism developed by economists.[94] The purpose of the EU ETS is "to promote reductions of greenhouse gases in a cost-effective and economically efficient manner".[95] In the EU ETS, emissions from the trading sectors are capped at set emission reduction levels and a corresponding amount of emission allowances are created and allocated to the market. By creating transferable pollution rights, optimal allocation of the pollution rights is expected to occur through the market, meaning that individuals valuing them the highest acquire the rights and reduction takes place where the marginal costs are the lowest. Consequently, the right to pollute becomes a production factor, like fuel or raw material. Negative externalities of pollution, for example climate change caused by GHG emissions, are internalised in the production costs. This corresponds with the polluter pays principle, recognised by the EU.[96] The transferability and allocation of pollution rights through the market sets the mechanism apart from traditional command-and-control environmental licenses.[97]

In addition to constituting an incentive-based steering mechanism, emissions trading contains elements and features of traditional administrative command-and-control regulation: The allowances are created and partially allocated to the market based on decisions adopted by national authorities and the Commission. The environmental integrity and effectiveness of the system, and thus the functioning of the market, is dependent on accurate information on emissions from GHG sources under its scope.[98] Hence, in addition to surrendering allowances, regulated entities are obliged to monitor, have verified and report GHG emissions on an annual basis in accordance with a pre-approved plan. Should they fail to fulfil these compliance obligations, they face an administrative fee determined by national authorities in addition to having their name published on a black list. National authorities supervise all of the compliance obligations.

Not all sectors and emission sources are automatically suitable for regulation through emissions trading. The regulatory choice of emissions trading as a steering mechanism and the extension of the scheme to new sectors has so far been made against the backdrop of the fulfilment of specific criteria. In conjunction with previous reforms of the EU ETS, the Commission has screened the suitability by imposing three requirements: 1) environmental effectiveness – in other words, the extent to which emissions trading can be expected to achieve the objective of reducing emissions; 2) cost-efficiency – the extent to which reductions can be achieved for a given level of resources/at the least cost; 3) consistency – the extent to which emissions trading is likely to limit trade-offs across the economic, social and environmental domain, and consistency with existing policies.[99] The

94 See e.g. R. H. Coase, *The Firm, the Market and the Law* (University of Chicago Press, 1990) and J. H. Dales, *Pollution, Property & Prices: An Essay in Policy-making and Economics* (2nd edn, University of Toronto Press, 1968), 58.

95 Consolidated Directive 2003/87/EC of The European Parliament and of the Council of 13 October 2003 establishing a scheme for greenhouse gas emission allowance trading within the Community and amending Council Directive 96/61/EC (the Emissions Trading Directive) [2008] OJ L 275/2003.

96 EU Commission 2019 (n 1).

97 See e.g. C. Streck and M. von Unger, 'Creating, Regulating and Allocating Rights to Offset and Pollute: Carbon Rights in Practice' (2016) CCLR 178, at 179.

98 EU Commission, Accompanying document to the Proposal for a Directive of the European Parliament and of the Council amending Directive 2003/87/EC so as to improve and extend the EU greenhouse gas emission allowance trading system – Impact Assessment [2007]SEC (2007) final 52, section 3.3.

99 EU Commission 2007 (n 99) sections 2.3 and 3.6.1.

Commission has assessed fulfilment of the aforementioned requirements based on specific criteria.[100]

The complete, consistent, transparent and accurate monitoring and reporting of GHGs, is considered fundamental by the EU for the effective operation of the ETS.[101] As the environmental integrity of the scheme relies on accurate emission information, the feasibility of acquiring reliable data on emission from emission sources is a key condition for the application of the EU ETS to a specific sector. Retrieving data on emissions from emission sources should be simple and the installation boundaries clear. On the other hand, the requirement of cost-efficiency poses limitations on the retrieving of emission data. In addition to the technical feasibility of MRV, the proportionality of transaction costs for participating in the scheme are evaluated. For small emission sources, for example cars, individual trucks and small ships, costs for MRV are high in relation to the emission reductions achieved and the administrative burden for managing the scheme quickly becomes high.[102] This is one of the reasons why, for example, road transport (consisting of a large amount of small emission sources) has not been included in the emissions trading scheme, despite the sector being considered on several occasions, starting from the establishment of the scheme.[103] For aviation, a point of regulation exists that fulfils the criteria of cost-efficient and accurate MRV of emissions; as aircraft operators constitute GHG sources that are sufficiently large to render emissions trading cost-efficient.[104] Certain small aircrafts benefit from a small emitters tool or a simplified verification procedure.[105] The EU ETS Directive outlines the principles and criteria applicable to MRV in the EU ETS,[106] and the Commission has adopted two implementing regulations, the Monitoring and Reporting Regulation (MRR) and the Verification and Accreditation Regulation, including detailed harmonised provisions on the MRV process. The monitoring and reporting of emissions reflects the most accurate and up-to-date scientific evidence available, in particular from the Inter-governmental Panel on Climate Change (IPCC).[107]

The MRV Shipping Regulation has been designed with the requirements on information used for the purposes of the EU ETS in mind. The MRV Shipping Regulation reflects the gradual approach of the EU in first creating a robust MRV framework, which functions as a "prerequisite for any market-based measure, efficiency standard or other measure, whether applied at Union level or globally" at a later stage.[108] The EU MRV Shipping Regulation uses, for example, a simplified verification system similar to the one applied under the EU

---

100 Ibid., at section 3.3.

101 Commission Implementing Regulation (EU) 2018/2066 of 19 December 2018 on the monitoring and reporting of greenhouse gas emissions pursuant to Directive 2003/87/EC of the European Parliament and of the Council and amending Commission Regulation (EU) No. 601/2012 (MRR) [2018] OJ L 334/1, recital 2.

102 EU Commission 2007 (n 99) section 3.3.3.

103 EU Commission, Green Paper on emissions trading within the European Union [2000] COM (2000) 87 final, section 10.

104 Ron CN Wit et al., "Giving Wings to Emission Trading – Inclusion of aviation under the European emission trading system (ETS): Design and Impacts" (Delft CE 2005), 52–53.

105 Regulation (EU) No. 421/2014 of the European Parliament and of the Council of 16 April 2014 amending Directive 2003/87/EC establishing a scheme for greenhouse gas emission allowance trading within the Community, in view of the implementation by 2020 of an international agreement applying a single global market-based measure to international aviation emissions [2014] OJ L129/1, art. 1.

106 Emissions Trading Directive (n 96), art. 14, Annex IV and Annex V.

107 Ibid., art. 14(2).

108 EU MRV Shipping Regulation (n 22), recital.

ETS, which is based on internationally agreed ISO standards and EU specific verification rules.[109] The EU MRV Shipping Regulation demonstrates that for shipping, MRV that fulfils the requirements of the EU ETS is possible. An inclusion in the EU ETS will pose specific requirements on the information collected and process for reporting and verification under the EU MRV Framework in the level of accuracy on the information as this is one of fundamental requirements of including shipping in the EU ETS.

12.5.2 The IMO process

Creating a CAP and TRADE system for emissions from transport, as suggested by the Commission, is based on giving carbon a price. The price can be generated through *emissions trading*, but it can also be generated through a *carbon levy*. A carbon levy means that an explicit price is placed on CO_2, or alternatively imposed through other costs that imply a carbon price. Under a carbon levy, the cost of controlling emissions would be certain (it would be equal to the levy), but, opposite to a CAP and TRADE system, there is no predetermined limit on emissions; hence the overall volume of emissions will remain unknown. The levy can be adjusted over time, but rather because of technical criteria or political considerations, than by the supply and demand of emission allowances. An important feature with a carbon levy is that funds can be collected and be distributed, for instance towards research and development.

Both systems are based on economic research, which show that a price tag on carbon will provide *information* on negative impacts of activities and hence incentivise behavioural change in relation to investments, production, and consumption patterns and promote technological development.[110] It will, in other words, work as a *nudge* towards emission reduction. Carbon pricing regulation affects also non-economic aspects of decision-making through the perceived reputational risks of being associated with high-emitting clients.[111] Case studies in the financial sector demonstrate that *carbon pricing* has a de-incentivising effect on financial support for high-emitting industries and projects already based on the information provided on emissions. As opposed to the EU, the IMO is not preparing for a CAP and TRADE system to reduce emissions from international shipping. While IMO remains committed to reducing GHG emissions from international shipping and, as a matter of urgency, aims to phase them out as soon as possible in this century, technological innovation and the global introduction of alternative fuels and/or energy sources for international shipping are considered integral to achieve the overall ambition.[112]

The latest step in the IMO emission reduction plan was taken by the Intersessional Working Group on Reduction of GHG Emissions from Ships (ISWG-GHG 7) in November

109 EU Commission, Impact Assessment Accompanying the document Proposal for a Regulation of the European Parliament and of the Council Amending Regulation (EU) 2015/757 in order to take appropriate account of the global data collection system for ship fuel oil consumption data [2019] SWD(2019) 10 final section 6.5.

110 High-Level Commission on Carbon Prices, "Report of the High-level Commission on Carbon Prices" (World Bank, 2017), section 3.3.

111 Megan Bowman, *Banking on Climate Change: How Finance Actors and Transnational Regulatory Regimes are Responding* (International Banking and Finance Law Series, Kluwer Law International 2015) 187.

112 www.imo.org/en/MediaCentre/HotTopics/Pages/Reducing-greenhouse-gas-emissions-from-ships.aspx (accessed 31.12.2020).

2020.[113] The group proposed draft amendments to the energy efficiency measures in MARPOL Annex VI chapter 4, building on the existing EEDI and SSEMP measures. According to the proposal, requirements to assess and measure the energy efficiency should apply to *all ships*, including existing vessels. Accordingly, two new measures were proposed: 1) Technical requirements to reduce carbon intensity, based on a new Energy Efficiency Existing Ship Index (EEXI); and 2) Operational carbon intensity reduction requirements, based on a new operational carbon intensity indicator (CII). The dual approach aims to address both technical (how the ship is retrofitted and equipped) and operational measures (how the ship operates). The proposed EEXI is required to be calculated for every ship of 5,000 gross tonnage and above (equal to the ships that are subject to the CMI DCS). These ships should also have determined their required annual operational CII. The CII determines the annual reduction factor needed to ensure continuous improvement of the ship's operational carbon intensity within a specific rating level,[114] which should be recorded in the ship's SEEMP.

According to the IMO framework, all large vessels (5,000 gross tonnage) new and old, are (or will be) under an obligation to collect and report on their fuel consumption and to apply to certain energy efficiency standards, all in order to comply with the UN development goals for emission reduction. With such systems available, the IMO will be equipped to enforce a global carbon levy for international shipping and by this finance research and development to Indorsee the technological innovation and the global introduction of alternative fuels and/or energy sources, which are deemed needed to reach the end goal of emission reduction. While committing to the same policy goal, the two organisations are developing different strategies, collecting information is a core element in both, but the collected information will be exploited differently.

12.6 Conclusions

Information has long been recognised as a policy instrument and utilised by policy makers on multiple levels. Furthermore, other policy instruments, such as emissions trading and carbon levy, rely on collected information; partly on GHG emissions and partly on fuel consumption. The purpose of this chapter has been to explore how information is utilised by policy makers and regulators (here the EU and the IMO) to achieve set policy goals on GHG emission reductions in international shipping.

Whereas the EU is collecting and distributing information on GHG emissions for vessels entering or leaving a port in the EU under the MRV Shipping Regulation, IMO is gathering information on fuel consumption. It is likely that both organisations will utilise the information in a market-based system with the intention to reduce emissions from shipping. In the EU an EU ETS for shipping is in the pipeline. Emissions trading as a policy instrument

113 The proposed amendments were made in an ISWG-GHG 7 remote meeting 19–23 October 2020. The draft amendments were then forwarded to the Marine Environment Protection Committee (MEPC 75), remote session 16–20 November 2020.

114 The rating would be given on a scale – operational carbon intensity rating A, B, C, D or E – indicating a major superior, minor superior, moderate, minor inferior, or inferior performance level. A ship rated D for three consecutive years, or E, would have to submit a corrective action plan, to show how the required index (C or above) would be achieved. Administrations, port authorities and other stakeholders as appropriate, are encouraged to provide incentives to ships rated as A or B.

itself does not function properly without accurate data on emission from sources but the information collected and published under it has the potential to work as a regulation instrument itself. The EU policy on emission reduction in transport, including shipping, is built on this insight. IMO on the other hand has so far not been willing to CAP emissions from international shipping. However, a carbon levy that would partly be utilised to reduce the use of fuels and partly to finance research and development on green innovations, which is considered necessary to achieve the needed reduction of CO_2 emissions, is discussed as an alternative to the proposed EU ETS. Most likely we have not yet seen the end of this discussion and other options will be studied as well. One insight remains clear: Business as usual cannot continue and international shipping will face a new regulatory framework with the intention to combat climate change in the near future.

CHAPTER 13

Liability for climate damage and shipping

*Ewan McGaughey**

13.1 Introduction

Climate damage, caused by burning coal, oil and gas, is the existential challenge of the 21st century, and the greatest threat to life on Earth since the end of the cold war. It is reversible, because nearly all uses of fossil fuels can already be replaced at lower financial cost (and far lower environmental cost) on our current state of technology. Wind, solar and batteries are now cheaper than coal, oil or gas-fired power stations.[1] The total cost of vehicle ownership for most kinds of land transport is already cheaper for electric than for petrol or diesel.[2] Shipping, however, presents a particular challenge, since the technology to develop heavy cargo and tankers is still in development. So how can the global shipping industry be helped to play its part in the fight to stop climate damage?

Global shipping accounts for up to 90% of world trade, and up to 4% of global greenhouse gas emissions. Since the industry is heavily concentrated, just four EU corporations account for over half of its market share, and therefore up to 2% of global emissions. Emissions such as carbon dioxide or methane trap more heat in the Earth's atmosphere, much like building a toxic greenhouse. This fact has been well understood since 1965 by the US government, since at least 1977 by Exxon Mobil, and is understood by shipping companies today. Greenhouse gas emissions cause average global temperatures to rise, and they have already done by over 1 degree since pre-industrial times. This causes more extreme weather events, such as fires from Australia, to the Arctic, to the Amazon, hurricanes in the Americas, or floods in Britain and Europe. To give just one example, the fires in Australia up to 7 January 2020 led to $700 million Australian dollars in insurance claims,[3] while

* Reader, School of Law, King's College, London. Research Associate, Centre for Business Research, University of Cambridge. I am very grateful to Andrew Tettenborn and Colm McGrath for comments. Please contact ewan.mcgaughey@kcl.ac.uk or @ewanmcg with all suggestions.

1 See e.g. M O'Boyle, "The Coal Cost Crossover: Economic Viability of Existing Coal Compared to New Local Wind and Solar Resources" (March 2019) Energy Innovation at https://energyinnovation.org/wp-content/uploads/2019/03/Coal-Cost-Crossover_Energy-Innovation_VCE_FINAL.pdf (accessed 22 January 2021). N Harmsen, "South Australia's Giant Tesla Battery Output and Storage Set to Increase by 50 Per Cent" (19 November 2019) ABC at www.abc.net.au/news/2019-11-19/sa-big-battery-set-to-get-even-bigger/11716784 (accessed 22 January 2021).

2 See e.g. K Palmer, JE Tait, Z Wadud and J Nellthorp, "Total Cost of Ownership and Market Share for Hybrid and Electric Vehicles in the UK, US and Japan" (2018) 209 Applied Energy 108, finding lower costs after 16,000 miles for hybrid vehicles, and that with government subsidies at the time of writing, electric cars were cheaper overall.

3 "Insurance Claims from Australia's Bushfire Crisis Exceed $700 million" (7 January 2020) SBS at www.sbs.com.au/news/insurance-claims-from-australia-s-bushfire-crisis-exceed-700-million (accessed 22 January 2022).

 DOI: 10.4324/9781003155195-13

AccuWeather estimated that the *total* economic loss from 2019 into 2020 would be $110 billion.[4] By burning fossil fuels on massive scale, the shipping industry is responsible for significant damage to the climate, and devastating loss of property and life. Yet the shipping industry continues to burn fossil fuels, while making profit, but not bearing the costs of its production. In economics this is an "externality". In law an orthodox analysis is that this is negligent or intentional harm.

This article examines how liability for climate damage may encourage rapid transition, and the elimination of fossil fuels, as fast as technologically possible in the shipping industry. Section 13.2 sets out the scientific, market, regulatory and technological background to shipping's role in climate damage. Section 13.3 examines corporate liability for climate damage, through the lens of emerging climate litigation cases. It discusses the challenges faced for corporate tort liability actions, either at common law or civil law, particularly in establishing a duty of care, breach, causation that is not too remote, and in the jurisprudence on policy and human rights. It analyses the leading cases which may inform the common law of tort, with reference to comparative litigation in *Lliuya v RWE AG* in Germany,[5] *Urgenda v Netherlands*[6] and *Smith v Fonterra Co-Operative Group Ltd* in New Zealand.[7] In essence, it is submitted that an orthodox application of principle suggests that greenhouse gas emitters owe a duty of care, are responsible for breach and cause damage that is not too remote. There is also, on orthodox application of principle, a duty of care on directors implicit in the Companies Act 2006 section 174 to adopt clean technology that will save a company money.[8] The argument that such actions are defective, a departure from principle, or "too big to work" are exaggerated.

Overshadowing the legal issues is a central issue of policy that needs to be confronted head on: is climate damage a phenomenon that is too big for a court? Is the control of multinational corporate power a task for the legislature alone? Even if liability would ensure that a polluter's externalisation of costs are internalised (a prediction that can itself be debated), are the implications for the distribution of resources not so great that this should be left exclusively to democratic decision? If we were to answer yes, this would be to concede that "the greater the damage, the less powerful the law". That conclusion cannot be accepted. The system of justice that remedies exploding dams, snails in ginger beer, or borstal boys and boats, is all in vain if it will not deal with the damage that eclipses all others. On the contrary, liability for climate damage is essential to ensure investments in infrastructure and technology are made, and to win the fight, particularly in shipping, for justice on a living planet.

4 J Roach, "Australia Wildfire Damages and Losses to Exceed $100 billion, AccuWeather estimates" (8 January 2020) AccuWeather including "damage to homes and businesses as well as their contents and cars, job and wage losses, farm and crop losses, infrastructure damage, auxiliary business losses, school closures and the costs of power outages to businesses and individuals as well as destruction of wildlife" At www.accuweather.com/en/business/australia-wildfire-economic-damages-and-losses-to-reach-110-billion/657235#:~:text=AccuWeather%20estimates%20the%20total%20damage,on%20a%20variety%20of%20sources. (accessed 22 January 2021).

5 (2015) Case No. 2 O 285/15 Essen Oberlandesgericht (holding that a claimant was entitled to establish causation between a German energy company's greenhouse gas emissions, and the damage to a village from melting of a Peruvian glacier).

6 (20 December 2019) 19/00135 (holding that the Dutch state had to reduce greenhouse gas emissions by 25% to protect the right to life under ECHR art. 2).

7 [2020] NZHC 419 (rejecting a claim in negligence against local greenhouse gas emitters).

8 By analogy to the duty of care in equity: *Medforth v Blake* [1999] EWCA Civ 1482, per Sir Richard Scott VC.

13.2 Science, shipping, law and technology

Before analysing the law of tort and its relation to shipping, it first makes sense to outline some essential background on (1) the causes of climate damage and science, (2) the implications as they relate to shipping, (3) the present state of regulation and (4) the technological possibilities for clean shipping.

13.2.1 Climate damage causes and science

The basics of climate damage are well known in the 21st century. Burning coal, oil and gas releases carbon dioxide, methane, nitrous oxide and other poisons into the atmosphere. If we want a living planet, it all must stop as fast as technologically possible. This simple fact is not often recognised squarely in public debate. Debate over what to do is habitually muddied by narratives of long-term targets, by language of climate change "mitigation", not reversal, and by talk of regulation or taxation of emissions, not bans. However, it is clear from historical examples, such as the ban on nuclear testing, the ban on aggressive war or the ban on slavery, that public narratives of "banning" are more effective at creating political coalitions than the language of regulation. They create moral clarity, which galvanises action.[9] This is also the language of the most effective international environmental law to date, the Montreal Protocol of 1987. This banned chlorofluorocarbons (CFCs) and hydrofluorocarbons (HFCs) and it has worked to close the hole in the Earth's ozone layer. Bans are also central to the language of tort law, which requires compensation or injunctions for harm. The ultimate cause of climate damage is therefore the failure of human institutions, and of our laws, to ban it. It is in this collective sense that we are all responsible.

The scientific cause of climate damage is the change in the composition of our atmosphere. Our air is made up of 78% Nitrogen (N_2), 20.9% oxygen (O_2), 0.934% argon (Ar), 0.0415% carbon dioxide (CO_2), and then neon, helium, methane (CH_4), krypton, nitrous oxide (N_2O) and many other gases. Carbon dioxide, methane and nitrous oxide particularly have the effect of trapping heat in the Earth's atmosphere as it radiates from the sun (nitrous oxide also directly harms human health). The right balance in this "greenhouse effect" is essential for life as we know it. With too little, we enter into an ice age. With too much, we become an uninhabitable furnace. The 0.0415% of carbon dioxide can also be expressed as 415 parts per million (ppm) of the air. This has been rapidly rising. In 1780, carbon dioxide was 0.0280% or 280ppm, it was 315 ppm in 1950, and 363ppm in 1997 as the Kyoto Protocol to the UN Framework Convention on Climate Change was signed.[10] Yet the science of climate damage was already understood in a 1965 report to US President Johnson, which said greenhouse gases would create "marked changes in climate" and this could be "deleterious from the point of view of human beings".[11] In 1977, Exxon Corporation had extensive internal research showing that "mankind is influencing the global climate through

9 F Green, "Anti-fossil Fuel Norms'"(2018) 150 Climatic Change 103.

10 The Kyoto Protocol committed countries to reduce six greenhouse gases in Annex A, namely carbon dioxide (CO_2), methane (CH_4), nitrous oxide (N_2O), hydrofluorocarbons (HFCs), perfluorocarbons (PFCs) and sulphur hexafluoride (SF_6).

11 R Revelle, W Broeker, H Craig, CD Keeling and J Smagorinsky, "Atmospheric Carbon Dioxide" in *Restoring the Quality of Our Environment: Report of the Environmental Pollution Panel* (November 1965) White House, President's Scientific Advisory Committee.

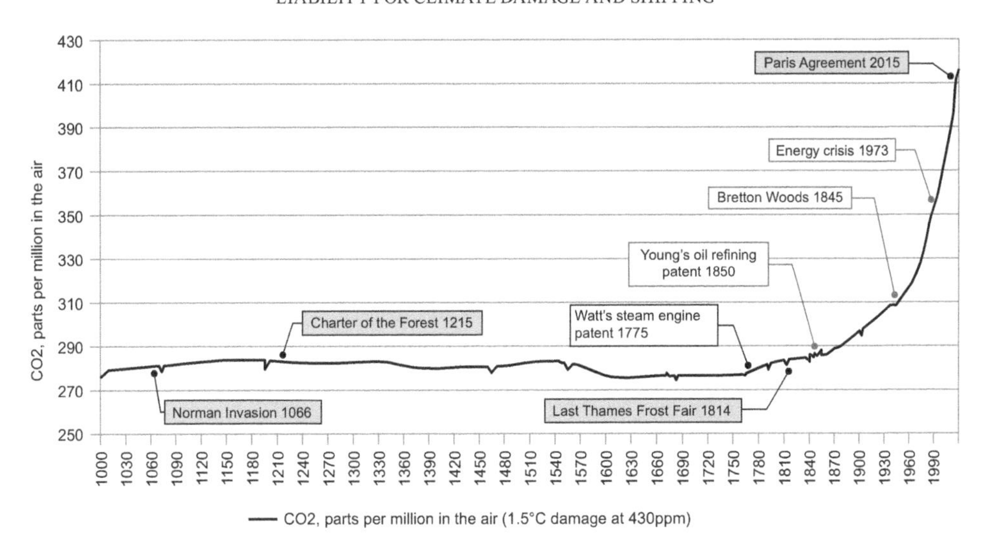

carbon dioxide release from the burning of fossil fuels", and in 1978 it said there was a "time window of five to ten years before the need for hard decisions regarding changes in energy strategies might become critical". Instead, Exxon worked to promote doubt about science,[12] a practice which is now systematic among fossil fuel creators and emitters and political parties captured by corporate money.[13]

Despite the exponential and existential threat of climate damage, the Kyoto Protocol's model in 1997 was that countries should set emission reduction targets,[14] without explicitly stating the need for fossil fuels to be shut down. In the Copenhagen Accord of 2009 major countries said there was (apparently) a "scientific view that the increase in global temperature should be below 2 degrees".[15] In the Paris Agreement of 2015, a further "ambition" of limiting temperature rises to 1.5°C was agreed.[16] However, the Paris Agreement, like all others before it, failed to simply state the problem – or even use the words – burning "coal", "oil" and "gas". In 2012, it was estimated that 565 more gigatons of carbon dioxide could be released into the atmosphere to stay below 2 degrees warming; 31 gigatons was released in 2011, and that limit was predicted to be reached in 2028. At that time fossil fuel companies had 2,795 gigatons worth of carbon in proven coal, oil and gas reserves, which would result in a 6–8 degrees rise.[17]

12 N Banerjee, L Song and D Hasemyer, "Exxon's Own Research Confirmed Fossil Fuels' Role in Global Warming Decades Ago" (16 September 2016) Inside Climate News https://insideclimatenews.org/news/16092015/exxons-own-research-confirmed-fossil-fuels-role-in-global-warming/ (accessed 22 January 2021).

13 See e.g. O Burkeman, "Memo Exposes Bush's New Green Strategy" (4 March 2003) Guardian, preferring the term "climate change" to "global warming", as it sounds more benign at www.theguardian.com/environment/2003/mar/04/usnews.climatechange (accessed 22 January 2021).

14 Kyoto Protocol of 1997 Annex B lists reductions compared to a base year, usually 1990. The UK was 92%.

15 The Copenhagen Accord (18 December 2009) is not a legally binding agreement, but a statement of intent.

16 This is a legally binding agreement in international law. Enforcement turns on domestic courts and pressure.

17 B McKibben, "Global Warming's Terrifying New Math" (19 July 2012) Rolling Stone at www.rollingstone.com/politics/politics-news/global-warmings-terrifying-new-math-188550/ (accessed 22 January 2021).

13.2.2 Global emissions shipping and market

The role of the global shipping industry matters because it both moves and burns fossil fuels. Between 80% and 90% of all world trade is carried out through the shipping industry.[18] Trade's geopolitical importance cannot be overstated, because economic integration contributes to international peace and stability. However, at the same time, over 40% of all shipping cargo is coal, oil and gas.[19] These are the very fossil fuels which cause climate damage, which is stated by the US Department of Defense to be an "'urgent and growing threat to our national security, contributing to increased disasters, refugee flows, and conflicts over basic resources such as food and water . . . projected to increase over time".[20]

Calculations of the global shipping industry's emission contribution range between 2.5%, suggested by the International Maritime Organisation,[21] and 3.3%, or 4% suggested by independent research: "just 15 of the world's biggest ships may now emit as much pollution as all the world's 760m cars".[22] Global shipping is concentrated, with the top 10 holding over 80% of the world market share.[23]

Table 13.1 Top 10 global shipping corporations, market share, and profit

	Corporation	*Countries*	*Ships*	*Market %*	*Profit 2019*
1	APM-Maersk	Denmark	681	17%	$5.5 bn
2	Mediterranean Shipping Co	Switzerland, Italy	574	16%	$1.4 bn
3	COSCO Group	China	487	12.4%	$5.5 bn
4	CMA CGM Group	France	533	11.8%	–
5	Hapag-Lloyd	Germany	238	7.2%	€373m
6	Ocean Network Express	Japan	210	6.5%	$105m
7	Evergreen Marine	Taiwan	197	5.3%	–
8	Hyundai Merchant Marine	South Korea	69	2.7%	–
9	Yang Ming Marine Transport Co	Taiwan	91	2.6%	–
10	Pacific International Lines	Singapore	108	1.4%	–

18 UNCTAD, *Review of Maritime Transport 2018* (2018) 23, suggesting 80%. IMO Profile, Overview (2020) suggesting 90% https://www.imo.org//en/OurWork/Environment/Pages/Default.aspx (accessed October 2021).

19 Maritime Strategies Research for the European Climate Foundation ***. C Davy, "Global Shipping Industry at Risk of Asset Stranding as Fossil Fuels Phased Out" (17 July 2019) China Dialogue Ocean at https://chinadialogueocean.net/9192-global-shipping-industry-asset-stranding/ (accessed 22 January 2021).

20 Department of Defense, *2015 Congressional Report on National Implications of Climate Change* (23 July 2015) 3. n.b, the Department of Defense itself releases as much emissions as 257 million cars.

21 IMO, Third IMO GHG Study 2014 (2014) for year 2012. EU, "Reducing Emissions from the Shipping Sector", ec.europa.eu. https://ec.europa.eu/clima/policies/transport/shipping_en (accessed 22 January 2021).

22 P Crist, "Greenhouse Gas Emissions Reduction Potential from International Shipping" (2009) OECD/ITF Working Paper 2009-11 at https://ideas.repec.org/p/oec/itfaaa/2009-11-en.html (accessed 22 January 2021). J Vidal, "Health Risks of Shipping Pollution Have Been 'Underestimated'" (9 April 2009) Guardian, citing a study by the Danish Environment Agency at www.theguardian.com/environment/2009/apr/09/shipping-pollution (accessed 22 January 2021).

23 Alphaliner, *Top 100* (2020).

As this table shows, the top four dominant European firms (Maersk, MSO, CMA and Hapag-Lloyd) hold 52.4% of global market share. It follows these big four European corporations are probably responsible for around 1% to 2% of global greenhouse gas emissions.

13.2.3 Current regulation of shipping

The current regulation of shipping with regard to greenhouse gas emissions is not satisfactory. Since 2012, the International Maritime Organization, based in London, has had an "ambition" to decrease shipping's greenhouse gas emissions "by at least 50% by 2050 compared to 2008 whilst pursuing efforts towards phasing them out".[24] This is problematic because, as explained above, when coal, oil and gas is eliminated from global trade, to be replaced by domestic wind and solar, shipping volumes will fall by 40% relative to current levels. This means the global shipping industry has the "ambition" to do much less than it seems, by a distant date, and even this is not (absent enforcement in tort)[25] a legally binding rule.

Within EU law, the Monitoring, Reporting and Verification Regulation 2015/757 does establish a system for monitoring, reporting and verifying carbon dioxide emissions, but there is little else. As recital 2 points out, other shipping emissions include nitrogen oxides, sulphur oxides, methane, particulate matter and black carbon. Yet only carbon dioxide emissions need to be reported, and only for ships over 5,000 tonnes.[26] In international law, the MARPOL Convention of 1973 initially focused on water waste and oil pollution, and this was updated in 2005 and 2020 to set loose limits on fuel efficiency and burning sulphur.[27] These changes will still mean that the shipping industry is responsible for killing around 250,000 people a year through air pollution, causing 6.4 million childhood asthma cases, and increasing greenhouse emissions as a result of sulphur reduction.[28] As well as ongoing, and direct damage to people's lives, the current regulation of shipping is not adequate to stop and reverse climate damage.

13.2.4 Technological possibility

What are the technological possibilities for a cleaner shipping industry? A first opportunity is simply the managed reduction of cargo volume, as countries shut down their use and reliance on coal, oil and gas. This will reduce total shipping loads by 40%, a fact that need not threaten profitability so long as shipping corporations and shareholders can expect and manage their strategic consolidation. A second opportunity is in developing solar-,

24 IMO, *Note by the International Maritime Organization to the UNFCCC Talanoa Dialogue* (2018) 3.1.3 at https://unfccc.int/sites/default/files/resource/250_IMO%20submission_Talanoa%20Dialogue_April%202018.pdf (accessed 22 January 2021).

25 Cf. *Sutradhar v Natural Environment Research Council* [2006] UKHL 33 (holding a geological report by the Natural Environment Research Council, which allegedly led Bangladeshi health authorities to not clean drinking water, could not ground liability).

26 MRV Regulation 2015/757 recital 2, and art. 2.

27 International Convention for the Prevention of Pollution from Ships of 1973, Annex VI. In the UK, there is power to give effect to this by Order from Merchant Shipping (Pollution) Act 2006 s 2, inserting Merchant Shipping Act 1995 s 128(1)(da).

28 M Sofiev, et al., "Cleaner Fuels for Ships Provide Public Health Benefits with Climate Tradeoffs" (2018) 9 Nature Communications 406 at www.nature.com/articles/s41467-017-02774-9 (accessed 22 January 2021).

hydrogen-, electric- or wind-based alternatives to fossil fuels on ships. On current research and development, solar power can only make up a fraction of power needed for mass cargo, for instance between 8% and 20% on one recent model.[29] Hydrogen power appears to work for a yacht, but though a recent project was sponsored by CMA CGM, it is still in a speculative state.[30] Concepts for a return to wind power, from massive sails or kites could be an option.[31] Yet the fact is that insignificant investment is being made in research and development at the time of writing, particularly compared to the billions in profit that the shipping industry makes. A central reason that shipping corporations feel no urgency is that they are currently bearing no cost for the damage their cargo and fuel use causes.

13.3 Corporate liability for climate damage

So far it has been explained how the global shipping industry contributes between 2.5% and 4% of global greenhouse gas emissions, that four European firms (Maersk, MSC, CMA and Hapag-Lloyd) are responsible for around half, and they calculate to make profit, despite knowing the damage they cause. It follows, in the language of economics, that shipping firms create "externalities", which make shipping a classic market failure.[32] Shipping prices are lower because companies are not paying for the damage they cause. This is a regulatory subsidy for the profits of a small group of global corporations. In the language of law, the facts that (1) burning fossil fuels foreseeably and directly damage the climate, (2) shipping companies fail to take reasonable care, (3) they increase chances of death and property destruction, and (4) the results are not remote but obvious, means that shipping companies either negligently, or if not intentionally, cause harm. On the face of it, this enables a clear set of actions in tort or crime.

Why have there not yet been cases for damage brought against shipping corporations – or indeed any major producer and emitter of greenhouse gas emissions – in UK courts? According to Professor Maria Lee, "any action against emitters will face profound and extensive doctrinal challenges, at every step of the process".[33] This is obviously true, as the table at part 2(2) also shows, because the *status quo* will continue to "produce large profits for powerful business interests".[34] Thus, it is necessary to identify the appropriate questions "at every step of the process". The following sections address three main groups

29 See e.g. M Gallucci, "EnergySails Aim to Harness Wind and Sun To Clean Up Cargo Ships" (14 April 2020) IEEE Spectrum at https://spectrum.ieee.org/energywise/energy/renewables/energysails-harness-wind-sun-clean-up-cargo-ships (accessed 22 January 2021).

30 See e.g. "Hydrogen: The Renewable Energy of Tomorrow's Shipping Industry' Speaking of its 'Ambition'" at www.cmacgm-group.com/en/Hydrogen-The-renewable-energy-of-tomorrow-s-shipping-industry#:~:text=When%20the%20boat%20sails%2C%20the,electricity%20needed%20by%20the%20ship (accessed 22 January 2021).

31 See e.g. Raunek, "Top 7 Green Ship Concepts Using Wind Energy" (7 October 2019) Marine Insight at www.marineinsight.com/green-shipping/top-7-green-ship-concepts-using-wind-energy/ (accessed 22 January 2021).

32 See JE Stiglitz and J Rosengard, *Economics of the Public Sector* ch 6. W.W. Norton & Company, Inc (New York) (2015).

33 M Lee, 'Climate Change Tort' (2015) ssrn.com, disapproving of overcoming the problem 'creatively' at https://papers.ssrn.com/sol3/papers.cfm?abstract_id=2695107 (accessed 22 January 2021).

34 *R (Sainsbury's Supermarkets Ltd) v Wolverhampton CC* [2010] UKSC 20, [81] per Lord Walker, referring to compulsory purchase rights for private companies. Cf. *R (Corner House Research) v Director of the Serious Fraud Office* [2008] UKHL 60, [55] per Baroness Hale referring to "the principle that no-one, including powerful British companies who do business for powerful foreign countries, is above the law".

of question in the tort of negligence: (1) Is there legally actionable damage, a duty of care, and breach? (2) Is there causation, which is not too remote? (3) Is it sound policy for the courts to impose a duty in law?

13.3.1 Damage, duty of care and breach

First, as outlined above, it is clear that there is damage from the burning of fossil fuels. This occurs through well-documented and scientifically understood processes of air pollution, climate change and the resulting extreme weather events. The damage done from burning fossil fuels, with extreme weather, causing loss of property and life, even aside from personal health problems from air pollution, eclipses all other forms of damage in human society, aside from nuclear meltdown.[35]

Second, is there a duty of care? In the established categories of damage to property and person,[36] this depends on whether harm is reasonably foreseeable, there is proximity, and it is fair, just and reasonable to impose a duty.[37] As we have noted, climate damage was foreseen already in 1965 by the US government, and fossil fuel companies like Exxon were doing extensive research on the issue from 1977. In most schools there has been extensive education on global warming since the 1990s, how it is caused, and how to stop it. This suggests that the connection between fossil fuel burning, and its effects, is not merely foreseeable, not merely known, but intended. It is not merely reckless pursuit of a dangerous activity while closing one's eyes to the consequences, but seeing the consequences and acting anyway, out of pursuit of profit. Shipping corporations continue because they believe they will not be liable for the costs they impose on others.

In *Smith v Fonterra Co-Operative Group Ltd*, the High Court of New Zealand Auckland Registry challenged this view. Mr Smith claimed that damage to coastal property was contributed to by a subsidiary of Bathurst Resources group (the largest coal miner in New Zealand), an oil refinery (owned mainly by ExxonMobil, BP, and Z Energy), a power station owned by Genesis Energy LP (in turn held mainly by US asset managers such as Invesco, ALPS Advisors, JP Morgan, and Chickasaw)[38] and other emitters. Mr Smith did not attempt to bring evidence of these entities' contribution to climate damage, or set about detailed quantification of loss to him, but in any event Wylie J rejected every element of a negligence claim. For Wylie J, because climate damage would occur anyway, and because he said the "defendants' collective emissions are miniscule" (without evidence of what they were), "reasonable persons in the shoes of the defendants could not have foreseen the damage". Such damage was (apparently) "an unlikely or distant result of the defendants' emissions".[39] This is logically false. Indeed, the parent company of one of the defendants, ExxonMobil, actually did foresee climate damage in 1977.

35 NB Nuclear Installations Act 1965 ss 12 and 16 creates strict liability for nuclear power station operators up to a series of limits, for instance €1.2 billion. It is notable that such a regime has not been passed by Parliament for fossil fuels.

36 Cf. *Robinson v Chief Constable of West Yorkshire Police* [2018] UKSC 4.

37 *Caparo Industries plc v Dickman* [1990] UKHL 2, adopting the reasoning but not result of Bingham LJ in [1989] QB 653.

38 See CNN Business, "Top 10 Owners of Genesis Energy LP" at http://money.cnn.com/quote/shareholders/shareholders.html?symb=GEL&subView=institutional (accessed 22 January 2021).

39 [2020] NZHC 419, [82].

The judge also rejected that the defendants had a requisite degree of "proximity", on the view that this requires assessment of physical closeness, or being in some pre-existing relationship.[40] This analysis imposes unwarranted strictures from cases on psychological harm, omissions, public service duties or assumption of responsibility,[41] to the simple damage of property.[42] The third component of a duty of care, whether it is "fair, just and reasonable", is examined at 13.3.3 below.

The question following from an established duty is, is there a breach? If a reasonable person should know their action could cause harm, but fails "to take reasonable care", there is a breach of duty.[43] However, it is plainly not the case that fossil fuel emitters merely fail to take reasonable care. Rather, they continue harmful conduct with full intent, because this makes more profit than investing in and moving to clean alternatives. In other words, action causing climate damage and resulting loss of property and life meets not only the basic requirements of negligence, but rises to an intentional tort. This also meets the more stringent standard in *Rookes v Barnard* for cases imposing punitive damages, namely "those in which the Defendant's conduct has been calculated by him to make a profit for himself which may well exceed the compensation payable to the plaintiff".[44]

13.3.2 Causation and remoteness

The next group of questions focuses on whether fossil fuel emissions can be said to "cause" damage that is not too "remote". It is obvious that no single entity is responsible for all damage to the climate (on a "but for" test), but rather multiple parties contribute to the harm. But in fact, global responsibility is concentrated in a similar fashion to the shipping industry. Just 90 corporate entities are responsible for 63% of all historic emissions.[45] There is, perhaps, some limited uncertainty about the link between any given source of emissions and any particular case of resulting damage, because emissions combine into overall increases in average global temperature. But in the face of any such uncertainty, it was established in *McGhee v National Coal Board* that a tortfeasor is responsible for injuries deriving from the "failure to take a step which materially increased the risk" of damage.[46] Where a case is analogous to the essential fact pattern in *Fairchild v Glenhaven Funeral Services Ltd*, as an "incremental and analogical development", tortfeasors who contribute

40 [2020] NZHC 419, [89]-[96].

41 See e.g. *Stovin v Wise* [1996] AC 923.

42 *Michael v Chief Constable of South Wales Police* [2015] UKSC 2, [106] per Lord Toulson, quoting Lord Bridge in *Caparo* pointing out that "proximity" and "fairness" "were little more than labels to attach to features of situations which the law recognised as giving rise to a duty of care". Property damage is plainly an established situation giving rise to a duty of care.

43 *Donoghue v Stevenson* [1931] UKHL 3, per Lord Atkin.

44 *Rookes v Barnard* [1964] UKHL 1, per Lord Devlin. Cf. Criminal Damage Act 1971 ss 1 and 5 (a defence is consent to damage). *R v G* [2003] UKHL 50, [41], quoting the Draft Criminal Code Bill, clause 18(c) that a person acts recklessly with respect to "(i) a circumstance when he is aware of a risk that it exists or will exist; (ii) a result when he is aware of a risk that it will occur; and it is, in the circumstances known to him, unreasonable to take the risk".

45 R Heede, "Tracing Anthropogenic Carbon Dioxide and Methane Emissions to Fossil Fuel and Cement Producers, 1854–2010" (2014) 122 Climatic Change 229 at https://link.springer.com/article/10.1007/s10584-013-0986-y (accessed 22 January 2021).

46 *McGhee v National Coal Board* [1972] UKHL 7. In the US, see *Sindell v Abbott Laboratories* 26 Cal 3d 588 (1980).

to damage are jointly and severally liable.[47] The alternative theory from *Barker v Corup (UK) plc* is that they may have proportionate liability, a rule that matters particularly where entities have gone insolvent.[48]

It follows that an entity will be responsible to pay compensation (C) according to its contribution as a percentage of historic emissions (HE), for damage to property or life (D), multiplied by the likelihood (L) that a particular event resulted from global warming.[49] Thus the climate damage formula is:

$$C = HE \times D \times L$$

For instance, if Maersk contributes up to 0.5% of historic greenhouse gas emissions annually, its share of total historic emissions must then be ascertained. This must be multiplied by the damage it has caused to a particular claimant or group, for instance, the Australian home-owners whose property burned down in the 2019–2020 bushfires, and take the $700 million estimated cost of insurers (who would be subrogated to homeowner claims) after the bushfires.[50] Finally, that figure must be multiplied by the percentage likelihood that the given event resulted from climate damage (let us assume a 0.9 or 90% likelihood), to arrive at a full figure for compensation from material increase in risk. If we assume, for the sake of example, that Maersk's historic emissions are the same as its annual emissions, this means:

$$C = 0.005 \times \$700,000,000 \times 0.9$$

On these (hypothetical) figures, Maersk would be liable to compensate Australian insurers (and indirectly the home and property owners) $3,150,000. Other greenhouse gas emitters would also be liable for their relevant shares. Following *Fairchild*, they would, further, be jointly liable with fossil fuel miners and producers for the oil, coal or gas that they use, and each could seek contribution from the other accordingly.[51] Following *Barker*, solvent companies would be liable for their proportion, but not for the losses of insolvent companies.

Such an approach is likely to be relevant to the recent, and ongoing German case of *Lliuya v RWE AG*. The Upper State Court (*Oberlandesgericht*) of Essen held that a major energy company, RWE AG, could in principle be liable for damage to a village from a melting glacier in Peru. Mr Lliuya's village, Huaraz, is near Lake Palcacocha. This was

47 *Fairchild v Glenhaven Funeral Services Ltd* [2002] UKHL 22, [2] and [34] and see its extension in *Novartis Grimsby Ltd v Cookson* [2007] EWCA Civ 1261.

48 *Barker v Corus (UK) plc* [2006] UKHL 20 which appears to follow the reasoning in H Hansmann and R Kraakman, "Toward unlimited shareholder liability for corporate torts" (1991) 100 Yale Law Journal 1879 but see the Compensation Act 2006, where Parliament disapproved of this limitation. It may be prudent to distinguish joint liability of individuals (an original concern of Hansmann and Kraakman) from corporate entities, though *Barker* was in principle approved again in *International Energy v Zurich* [2015] UKSC 33.

49 Three distinctions are therefore between (1) annual emissions and all historic emissions: the latter is the relevant figure for computing compensation, (2) events such as fires, hurricanes or floods that become more likely, and ongoing damage such as glaciers melting that are entirely caused by global emissions.

50 This example assumes that insurers may be subrogated to the claims of their policyholders: see generally C Mitchell and S Wattterson, *Subrogation: Law and Practice* (OUP, 2007), ch.10.

51 An interesting question is, what should be the just contribution? If A gives B oil knowing it will be burned, a good starting presumption is equal liability. This is fundamentally different to consumers, who have no choice over the products available on the market.

increasing in size as a nearby glacier melted, requiring flood defences for the village. It was submitted that RWE had contributed to 0.47% of global carbon emissions, and is liable for 0.47% of the cost of flood protection. While the State Court dismissed the claim, arguing there was no "linear casual chain", in November 2017 the Upper State Court admitted the claim to move to the evidential phase on the threat of mudslides, and RWE's exact contribution, to be assessed by experts.[52]

The judgment of the Upper State Court in *Lliuya v RWE AG* contrasts with the judgment of Wylie J in *Smith v Fonterra*. This said establishing any causal contribution was "impossible", on the footing that the material increase in risk test as developed under *Fairchild* had been a move laden with "peril".[53] This is not true for three reasons. First, climate damage involves far less uncertainty than mesothelioma cases: whereas nobody knows whether any one employer is responsible for the "fatal fibres" of asbestos, we know that greenhouse gas emitters are responsible in roughly equal amounts for the increases in concentration of atmospheric toxins. Second, detailed methods of measurement and calculation are established both in scientific practice and in law to quantify emissions and their effects. In the case of shipping, an example is the Maritime Carbon Emissions Regulation 2015 (above at part 2(3)). Third, the material increase in risk test long predated *Fairchild*, and embodies a key point of principle: if there are more tortfeasors contributing to damage, all of them (not none of them) are responsible.

A related question is whether a causal chain leading from emissions to damage from climate change is too "remote". The test since *The Wagon Mound* or *Cambridge Water* is whether the resulting type of damage is reasonably foreseeable. If not, then the consequence will be too remote.[54] Climate damage is not a remote consequence of burning coal, oil and gas, because this has been understood and expected for over 50 years.

13.3.3 Policy and human rights

A third category of question overshadows, perhaps, all of the more technical steps in the common law framework of negligence. Should judges decide profoundly political issues? Is a climate damage claim "too big to work"? In negligence, this issue arises under whether it is "fair, just and reasonable" to impose a duty of care. This is similar to asking whether a claim is "justiciable", and ultimately whether it encompasses a sound policy.[55] This creates a clear set of choices for all courts.

A first option is to deny claims completely. In 2007, the US Supreme Court held that the Environmental Protection Agency had a duty to regulate greenhouse gas emissions under the Clean Air Act of 1963,[56] but this apparent victory for environmental groups was short lived, as *American Electric Power Co v Connecticut* held that given the EPA's

52 (2015) Case No. 2 O 285/15 Essen Oberlandesgericht.

53 [2020] NZHC 419, [88].

54 *Overseas Tankship (UK) Ltd v Morts Dock and Engineering Co Ltd* [1961] UKPC 2 (replacing the directness test) and *Cambridge Water Co Ltd v Eastern Counties Leather plc* [1993] UKHL 12 (remote when nobody knew chemicals seeped through concrete).

55 Cf. *Marcic v Thames Water plc* [2003] UKHL 66, [70] per Lord Hoffmann (a nuisance claim, rejecting that a public authority had a duty to build new sewers).

56 *Massachusetts v Environmental Protection Agency*, 549 US 497 (2007), under 42 USC §7521(a)(1).

jurisdiction, federal tort law and public nuisance,[57] was pre-empted.[58] For the US Supreme Court, the fact that statute "speak[s] directly to [the] question" displaced the claim against the five largest power companies, emitting 650 million tonnes of CO_2 a year, for contributing to global warming.[59] The Supreme Court majority did suggest that state tort law might not be pre-empted,[60] but in subsequent cases courts may conclude that, as the issue crosses state boundaries, state law is pre-empted as well.[61] This uniquely broad pre-emption doctrine ignores that environmental laws aim to set minimum, not fixed standards.[62] Similarly, in *Smith v Fonterra*, Wylie J argued the "most appropriately placed entity . . . to address the complex and collective problems presented by climate change is the Government". In his view, "liability in negligence would potentially compromise Parliament's response", illegitimately engaging courts in "complex polycentric issues" involving "policy formation, value judgments, risk analysis, trade-offs and distributional outcomes".[63] Mr Smith's claim was "too big to work". This option is an affront to a serious system of justice. Not making a decision is itself determining a clear distributional and political outcome, giving an unbridled, state-backed licence to damage to other people's property.

A second option is to interpret the law in light of human rights, and international law. In *Urgenda v State of Netherlands* the Dutch Supreme Court held that the Dutch government had to reduce greenhouse gas emissions by 25% before 2020, a standard taken from the UN Framework Convention on Climate Change, Annex I, and a report of the IPCC in 2007. This lists developed countries which must reduce emissions between 25% and 40% by the end of 2020 to stay below 2 degrees of climate damage.[64] The Dutch Supreme Court held failure to meet even the lower end of this target violated both the right to life in the European Convention on Human Rights article 2, and the right to private and family life in ECHR Article 8.[65] The significance of a decision on these grounds is that, if followed by other courts around Europe, UK courts have good reason to interpret common law norms, including the tort of negligence, so as to be compatible with the protection of the right to life and privacy.[66] This option has the advantage of ensuring that domestic law upholds, rather than undermines, international law. This is all the more important since global warming is an international, planetary problem.

---

57 While public nuisance was pleaded under US law, it appears that negligence is the more appropriate tort in English law.

58 *American Electric Power Co v Connecticut*, 564 US 410 (2011).

59 564 US 410 (2011) per Ginsburg J. See also *Native Village of Kivalina v ExxonMobil Corp* 696 F 3d 849 (2012) *certiorari* (i.e. appeal) denied by Supreme Court (2013), claimed damages for sea level rise and future floods.

60 564 US 410 (2011) Ginsburg J, leaving "the matter open for consideration on remand".

61 See e.g. *City of New York v BP plc* (2nd Cir 2020) is pending at the time of writing.

62 Cf. 42 USC §7401(a)(3) referring to the "primary responsibility of States and local programs", with (4) federal assistance. This does not refer to "exclusive responsibility".

63 [2020] NZHC 419, [92] and [98].

64 UN Framework Convention on Climate Change 1992, Annex I, and recommendation from the IPCC Working Group III, *Climate Change 2007: Mitigation of Climate Change* (2007) Contribution to the Fourth Assessment Report, ch 13, 776, Box 13.7 at www.ipcc.ch/site/assets/uploads/2018/03/ar4_wg2_full_report.pdf (accessed 22 January 201). Note that the Renewable Energy Directive 2009/28/EC Annex I target for the Netherlands was just 14%.

65 Judgment (20 December 2019) 19/00135, text in Dutch but best translated through deepl.com.

66 Human Rights Act 1998 s 6 and *R (Unison) v Lord Chancellor* [2017] UKSC 51, [89] per Lord Reed, ECHR case law is "relevant to the development of the common law".

A third approach is to apply established principles of liability that already exist at common law. Is harm reasonably foreseeable? Is there a failure to take reasonable care? Is there causation that is not too remote? For all of these questions, the answers are "yes", and they are "yes" despite the inevitably massive volumes of corporate money that will be thrown into every opportunity to delay action in litigation. It is true that the common law should be interpreted consistently with statute,[67] but there is simply no limit on liability for climate damage as it stands. In declaring slavery unlawful at common law, in 1772, Lord Mansfield memorably said "let justice be done whatever be the consequence".[68] The consequences of making shipping corporations and other fossil fuel polluters pay for the harm they cause will likely be (1) a fall in those corporations' traded share prices, to price in the costs of decarbonising production, and future compensation, (2) the insolvency of a minor number of companies that cannot diversify their business models, (3) the actual compensation of some people who have been harmed by climate damage, and (4) a response by Parliament to set a statutory framework for accelerated transition to clean energy, or possibly its limitation. These are consequences about which a court will be mindful, but by which no court should be swayed.

13.4 A director's duty to decarbonise

A related aspect of negligence liability is the duty of directors under the Companies Act 2006 section 174. While tort claims focus on damage to third party victims of a company's conduct, company members (and arguably other stakeholders such as bondholders, or beneficial investors)[69] may also bring derivative claims against directors for negligent loss to the company. This duty of "care, skill and diligence" is judged according to the director's subjective skills, added to a baseline of objective care that should be shown given the director's role and the company's size.[70] It raises two main types of claim. First, a director that fails to take action to cease carbon emissions, and subsequently faces tort claims, could be liable to compensate the company. Premiums for "Director & Officer" insurance will probably be affected if the likely cost of defending claims were to rise. This in itself could create sufficient pressure for pre-emptive action by companies to decarbonise production.

Second, directors could be liable for negligently failing to eliminate fossil fuel energy simply on the ground that fossil fuels cost more money. This flows from the general duty of care imposed upon those in a fiduciary position. For example, in *Medforth v Blake* the Court of Appeal held that a receiver and manager was liable to the owner of an insolvent farming business after he failed – despite being told by the owner – to get £1,000 per week in discounts for pig food. This resulted in excessive costs being run up, and therefore breached a general duty of "good faith", and within that "a duty to manage the property with due diligence", including "reasonable steps to be taken in order to try to do so profitably".[71]

67 See e.g. *Johnson v Unisys Ltd* [2001] UKHL 13.

68 *Somerset v Stewart* (1772) 98 ER 499.

69 See e.g. *Re Fort Gilkicker* [2013] EWHC 348 and *BCE Inc v 1976 Debentureholders* [2008] 3 SCR 560.

70 Companies Act 2006 s 174(2).

71 *Medforth v Blake* [1999] EWCA Civ 1482, per Sir Richard Scott VC.

The standards for insolvency practitioners in *Medforth v Blake* are applicable to company directors in the fulfilment of their duties,[72] since section 170(4) makes clear that duties must "be interpreted and applied in the same way as common law rules or equitable principles".[73] It follows that where directors could save money, for instance, by (1) buying electric vehicles, with lower total costs than fossil fuel vehicles,[74] (2) installing solar panels or building wind turbines to acquire energy more cheaply than from the grid,[75] or (3) insulating buildings to reduce energy bills, and they fail to take such simple steps, they must be regarded as liable for negligence under section 174, for the extra costs to the company.

How might this standard apply to directors of shipping corporations, where technological alternatives to oil powered tankers and cargo ships are limited? While general tort liability would encourage shipping firms to rapidly invest in clean technology, this director's duty would mean that where any cheaper alternative becomes available, it must be rapidly adopted.

13.5 Conclusions

The benefits of a cleaner shipping industry are tremendous, and they are necessary for the continued integration of the global economy, sustainable trade and a living planet. While current regulation and technology is inadequate, this chapter has explained how shipping corporations, and their directors, can be made liable for their contribution to climate damage. This is the best way to ensure that rapid steps are taken to shift to clean shipping, based on solar, wind, battery or other technology, and to ensure that the planet is healthy and habitable for generations to come.

72 It also applies to asset managers in making investment decisions, cf. *Harries v Church Commissioners of England and Wales* [1992] 1 WLR 1241, holding that asset managers must act in their clients' interests, including moral or financial interests.

73 Companies Act 2006 s 170(4).

74 Cf. K Palmer et al., "Total cost of ownership and market share for hybrid and electric vehicles in the UK, US and Japan" (2018) 209 Applied Energy 108.

75 See e.g. I Johnston, "Tesco to switch to 100% renewable electricity this year in UK and by 2030 worldwide" (16 May 2017) Independent at www.independent.co.uk/environment/tesco-switch-renewable-lectricity-energy-uk-2017-2030-worldwide-supermarket-a7739076.html (accessed 22 January 2021). "Volvo Cars unveils first solar energy installation at Ghent car factory" (4 October 2018) Automotive World. E Logiudice at www.media.volvocars.com/global/en-gb/media/pressreleases/238862/volvo-cars-unveils-first-solar-energy-installation-at-ghent-car-factory (accessed 22 January 2021). "Solar power is saving this farmer money" (2 October 2018) Yale Climate Connections at https://yaleclimateconnections.org/2018/10/solar-power-is-saving-this-farmer-money/ (accessed 22 January 2021).

INDEX

Note: Numbers in **bold** indicate a table